Praise for *How to Raise the*
Vibration around You

"*Earth, Water, Air and Fire (Light) are a means of representing the Universe in many traditions. The sequence describes the principle of interconnectedness and is the metaphor organizing this excellent book by Dawn James. These four elements provide the basis to understand our patterns of life and well-being. Dawn offers a practical set of guidelines to create harmony and balance within the self and in the work and home domains. From crystals to essential oils—and the common sense of our ancestors—eco-friendly direction is there in abundance. What resonated strongly with me as a reader was the notion of 'sound' as the vocabulary of Nature. Dawn uses a profound template to organize this work.*"

~ Dr. Ian Prattis, Professor Emeritus
www.ianprattis.com

"*Dawn James presents a practical and insightful roadmap on how to shift the vibrations around us. Her book How to Raise the Vibration around You provides a real-life understanding that everyone needs to help us anchor Heaven right here on Earth, and our homes and offices too!*"

~ Susan Kern, MSc
www.magicallifeinstitute.com

"*In an increasingly complex world, Dawn shows us how we can use simple and natural solutions to help keep our bodies, homes and work spaces healthy. I highly recommend this book for people who feel sick and tired. The tips and guides in the book can help anyone who's 'tried everything' actually find ways of healing that work!*"

~ Tash Jefferies, Television Host
www.tashjefferies.com
www.tvhealthy.com

How to Raise the Vibration around You

How to Raise the Vibration around You

Volume I:
Working with the 4 Elements to Create Healthy and Harmonious Living Spaces

Dawn James

Library and Archives Canada Cataloguing in Publication

James, Dawn, 1965-, author
    How to raise the vibration around you / Dawn James.

Contents: v. 1. Working with the 4 elements to create healthy and
    harmonious living spaces.
Issued in print and electronic formats.

ISBN 978-0-9865378-0-6 (pbk. : v. 1).—ISBN 978-0-9916715-3-3 (epub : v. 1).—
ISBN 978-0-9865378-4-4 (pdf : v. 1)

    1. Interior decoration—Environmental aspects. 2. Sustainable living.
3. Vibration. I. Title. II. Title: Working with the 4 elements to create healthy
and harmonious living spaces.

NK2113.J35 2014   747   C2013-905372-7
                                                          C2013-905373-5

Design and layout by Stacie Scherer/Pass It On Communications Inc.
Cover design by DDMPR Company
Edited by Andrea Lemieux

Lotus Moon Press
P. O. Box 1665
Brighton, Ontario
Canada K0K 1H0
Raiseyourvibration.ca

Printed and bound in Canada.

Dedication

How we treat the Earth is a reflection
of how we treat our bodies;

the condition of the Water is a reflection
of the condition of our blood;

what we put in the Air is returned to us
in a breath; and

when we shun the Light of the heavenly bodies,

our Spirit becomes entrapped in a calcified shell.

Take heed and reflect on these things.

They are gentle reminders of a promise
you made long ago.

Remember your promise to protect, care for
and cherish this planet.

Remember your promise to live in harmony with nature and
all living things.

Remember your promise to give as you receive,

and, above all, remember your promise
to be an Earth keeper.

This book is dedicated to those on the path of remembering
who they are, what they are and why they are here.

Contents

Part III: Water

Part IV: Earth's Gifts

Part V: The Ultimate Goal

Introduction

*N*ature provides us with a bounty of natural ways to maintain our health and wellness, and it is my intention to teach you how to recognize and work with these gifts of nature to create healthy and harmonious living spaces. When we are in harmony with nature and the elements, we naturally appreciate, cherish and protect this beautiful planet that all living things depend on.

In my first book, *Raise Your Vibration, Transform Your Life: A Practical Guide for Attaining Better Health, Vitality and Inner Peace*, we explored and expanded on our understanding of the mind-body-spirit connection and how one intelligent energy system (bioelectricity) strengthens this network, and, more importantly, what we can do to positively raise our personal frequency on the physical, emotional, mental and spiritual levels.

In this second book in my trilogy of understanding vibrational frequency, *How to Raise the Vibration around You: Volume I: Working with the 4 Elements to Create Healthy and Harmonious Living Spaces*, we explore the world of vibrational frequency beyond the physical self and discover numerous ways to raise the frequency in your home, at work and in your general surroundings by working with the four elements, air, light, water and Earth's gifts. The third book in the trilogy, volume II of *How to Raise the Vibration around You*, will address how to raise the vibration around us working with the fifth element—space.

This book is divided into four parts, each representing one of the four elements. Part I, "Air," explores several ways to increase the vibrational frequency of the spaces you occupy through the harmonic use of incense, essential oils, natural air purifiers and filters, wind, music and the sounds of nature.

Part II, "Light," explores how to increase the frequency in the spaces you occupy through the use of natural light, candles and indoor lighting, as well as the benefits of working with sunlight and moonlight.

Part III, "Water," explores several ways to increase the frequency in the spaces you occupy through the use and conditioning of water for cleaning, bathing, drinking and sensory enjoyment.

Part IV, "Earth's Gifts," explores several ways to increase the frequency in the spaces you occupy by using Earth's bounty of natural gifts, such as crystals and gemstones, and bringing nature indoors, as well as how to reconnect to the earth in a physical way. One or a combination of the methods presented in this book can promote health and create harmony in your living and work spaces.

And last, but certainly not least, we arrive at the ultimate goal: what it takes to be a harmonic being and enjoy peace and harmony in all aspects of life, regardless of your circumstances.

At the back of the book, you will find additional resources and a suggested reading list so you can continue the journey and share what you have learned with others.

Part I

∞

$\mathcal{A}$ir

Smell and the Olfactory System

The nose knows.
~ Anonymous

*O*ne of the ways we can raise the vibrational frequency around us is to use certain aromatics that release fragrance into the air. Why would we want to do this?

Our sense of smell is governed by the olfactory system, which involves the processing of information about the identity, concentration and quality of a wide range of chemical stimuli. These chemical stimuli—called odorants—travel up into the nose, activating the olfactory system, which then relays information to the brain.

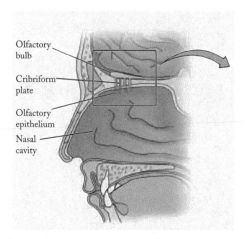

Peripheral and Central Components of the Olfactory Pathway[1]

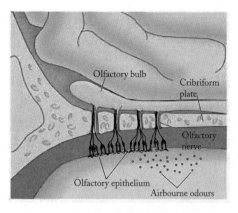

Receptor Neurons, Epithelium and Bulb of the Olfactory System[2]

What makes the olfactory system unique is that it projects information to a number of areas of the forebrain, including the hypothalamus, amygdala and cerebral cortex, which further stimulate various motor, visceral and emotional reactions.

 4

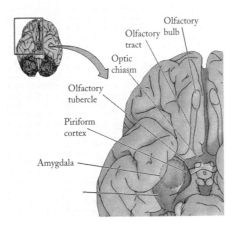

Olfactory
Olfactory bulb
tract
Optic
chiasm
Olfactory
tubercle
Piriform
cortex
Amygdala

**The Areas of the Forebrain That Are Stimulated
through Our Sense of Smell[3]**

Many of these areas of the brain are part of the limbic system, which is involved with emotional behavior and memory. That's why, when you smell something, it often brings back memories associated with the object.

In the following two sections—incense and essential oils—you will discover how adding certain fragrances to the air can alter the vibration of your surroundings and positively affect your well-being.

Incense

───────────────∞───────────────

Let my prayer be set forth before thee
as incense; and the lifting up of
my hands as the evening sacrifice.
~ Psalms 141:2

───────────────∞───────────────

The word incense comes from Latin *incendere*, which means "to burn." When incense burns, it releases a fragrant smoke. Incense is largely used for religious worship, creating a spiritual atmosphere in religious ceremonies. However, it is also used in aromatherapy and meditation, often combined with essential oils, and it can be used to mask unpleasant odors.

The History of Incense

Chinese cultures from the late Stone Age used incense. Records show that the ancient Chinese used incense from herbs and plants such as cassia, cinnamon, styrax and sandalwood in formal ceremonial rites. Later, the Hindus, who learned about incense from the Chinese, were the first to use the roots of plants. The Vedas, ancient Hindu texts, indicate that incense was likely used 6,000 to 8,500 years ago.

The ancient Egyptians used incense to camouflage bad odors as well as to drive away evil spirits and please the gods. The oldest incense burner ever found goes back to the fifth dynasty of Egypt (from about 2494 to 2345 BCE), and the Temple of Deir el-Bahari, built in the fifteenth century BC in Egypt, holds a series of carvings depicting an expedition for incense.

At around 2000 BCE, the Chinese were the first to use incense for religious worship, with usage peaking during the Song Dynasty (CE 960 to 1269), with many buildings being erected specifically for incense ceremonies.

Incense use also spread to Babylon, where it was used during prayers to oracles, and from Babylon, it spread to Greece and Rome. It also spread to Japan in the sixth century CE through Korean Buddhist monks, who used incense in their purification rights.

In fourteenth century CE Japan, a samurai warrior might perfume his armor with incense to project an aura of invincibility, and during the fifteenth and sixteenth centuries CE, the appreciation of incense spread to the Japanese middle and upper classes.

Types of Incense

There are two main types of incense: indirect burning and direct burning. Indirect-burning incense, or non-combustible incense, cannot burn on its own; it needs a separate heat source, traditionally provided by charcoal. Depending on the texture of the material used to make this type of incense, the duration of burning will vary. Powdered or granulated material burns more rapidly, whereas coarsely ground material or whole chunks burn more gradually. Often this type of incense is combined with a binder such as honey or a soft resin and formed into balls. Examples of indirect-burning incense are frankincense and myrrh.

Direct-burning incense, or combustible incense, can be lit directly by a flame that is then fanned out to create a smoldering, glowing ember that releases the fragrance of the incense. This incense is made from a moldable substrate of the incense material, combined with an odorless binder and shaped into sticks (joss sticks), cones, pyramids and other forms.

Uses of Incense
The uses of incense around the world are as diverse as the people who use it and their practices. Throughout history, incense has had practical, religious and aesthetic uses.

Practical Uses of Incense
Because of the strength of incense fragrances, it has been used to obscure the odor of decay in funerary ceremonies. Incense made from citronella and similar fragrances, often manufactured in coil form, is used to repel mosquitoes and other insects, and some smokers use incense to mask the tobacco odor that might linger indoors.

Religious Uses of Incense
Incense has had religious uses since antiquity. It can aid in prayer and be used as a sacrificial offering to a deity, which was common in Judaic worship. Its use is prevalent in many religions, despite their not having a lot in common. Eastern Christian and Roman Catholic churches use incense, often as a symbol of prayer rising to heaven. The use of incense as a form of sacrificial offering is still used in the Roman Catholic, Anglican and Orthodox churches, and in Taoist and Buddhist Chinese religions.

Aesthetic Use of Incense

Incense can be used simply for its perfume, with no particular significance being given to it.

Sixteen Common Incenses and Their Elemental Attributes

Amber: Amber is associated with the element Air. It is good for meditation when seeking information and for past-life regressions and divinations into the past.

Cedarwood: Cedar is associated with the element Fire, and it denotes strength and power.

Cinnamon: Cinnamon is associated with the element Fire. It is related to passion, lust and sexual desire and is very good for raising energy.

Frankincense: Frankincense is associated with the element Water, and it promotes peace and calm. Commonly used to relieve anxiety and stress, it is helpful in reestablishing friendships after conflicts and arguments, relieving the tension in a tense situation. Frankincense is also used for religious rituals and meditation.

Ginger: Ginger is associated with the element Fire, and it encourages desire or lust.

Jasmine: Jasmine is associated with the element Water, and to a lesser extent, Earth. Jasmine is known as an incense of the night and of love, though not of sex, and it is associated with inner beauty and friendship.

Lavender: Lavender is associated with the element Water. It promotes beauty and is often burned during romantic situations.

Lotus: Lotus is associated with the element Air and is related to mental clarity, increased focus and heightened intelligence. Lotus is ideal for meditation.

Musk: Musk is associated with the element Earth. This incense is used to restore balance and order because it realigns one's energies and seals openings, creating barriers and protection.

Myrrh: Myrrh is associated with the Spirit and the Sun because, like the sun, which renews itself each day, myrrh elevates one to the heavens and promotes renewal in the next life. For this reason, myrrh is used as a funerary incense. Because of this divine association, myrrh is often used for purification and exorcism.

Nag Champa: Nag champa is associated with all four elements—Air, Fire, Water and Earth, as well as Spirit. This sacred incense, which is blended with sandalwood, can be used for many of the same reasons as sandalwood. Nag champa is used to sanctify and purify an area, and it is a good general-purpose incense in spiritual situations, such as meditation and when seeking spiritual enlightenment or development.

Opium: Opium is associated with the element Air, and it is burned to induce lucid or prophetic dreams.

Patchouli: Patchouli is associated with Fire and Earth, and it inspires happiness and joy.

Rose: Rose is associated with the element Water, and to a lesser extent, Fire. Rose incense relates to love, sex and desire, and it is good to burn to inspire a more romantic situation.

Sandalwood: Sandalwood is associated with the elements Fire and Water. It is also a divine, or spiritual, wood and is used to purify and sanctify an area.

Vanilla: Vanilla is associated with the element Air and is good for heightening thought, concentration and reflection.

———————————————∞———————————————

Essential Oils

―――――――∞―――――――

For breath is life, and if you breathe
well you will live long on earth.
~ Sanskrit proverb

―――――――∞―――――――

Essential oils are the natural volatile oils that are extracted from the aromatic essences of certain plants, shrubs, flowers, trees and herbs. They contain virtually all of the plants' nutrients, and their vibrational frequency is compatible and beneficial to human vibrational frequency.

Essential oils (EOs) are not used internally, but are inhaled or applied to the skin. Clinical research shows that these concentrated oils have the highest vibrational frequency of any natural substance known to man. Furthermore, EOs produce the highest level of oxygenating molecules of any substance currently known to man.[4]

Why is this important for our health and well-being? Because bacteria cannot live in an oxygen-rich, high-vibrational environment!

One of the best ways to raise the vibrational frequency around you is to infuse the air with essential oils. This can be done with a diffuser, a cloth or tissue, hot water, a vaporizer or humidifier, or a fan or vent. One of the easiest ways to disperse oil in the air for inhalation is to use an aroma diffuser. A cool-air diffuser is recommended rather than a diffuser that uses heat, such as a light-bulb ring, because heat alters the chemical constituents of the oil, reducing its therapeutic properties.

Essential oil is put into the cool-air diffuser, and under the pressure of air blowing through a nebulizer, which reduces the oil to a fine mist, the oil is dispersed into the air, covering hundreds of square feet in minutes. The essential oils, with their oxygenating molecules, remain suspended for several hours and improve air quality.

Essential Oils as Air Cleaners

Essential oils have antiviral, antibacterial and antiseptic properties that have been found to help kill bacteria and reduce fungus and mold. They have also been found, when diffused, to reduce the amount of airborne chemicals and metallic substances, as well as uplift one's spirits and create emotional harmony. All you need to do is turn on a cool-air aroma diffuser for fifteen minutes each hour.

You can also put one to two drops of oil on a cotton ball and attach it to a ceiling fan or an air vent. This can also work well in a vehicle.

What types of essential oils to use is a matter of personal taste. The following chart can help you choose an oil or oils to suit your preferences and needs.

Essential Oils and Their
Aromatic Influences[5]

Essential Oil	Aromatic Influence
Angelica	Helps release and let go of negative feelings and anger
Basil	Helps maintain an open mind and clarity of thought
Bergamot	Uplifting and refreshing; may help open the heart chakra and reduce stress and anxiety
Cardamom	Invigorating, uplifting; may be beneficial for clearing confusion
Cedar, Western red	Calming; may help enhance spiritual awareness or meditation
Chamomile, Roman	Helps stabilize emotions, dispel anger and soothe and calm the mind, creating an atmosphere of peace and patience
Clove	May create feelings of protection and courage; a mental stimulant that may improve memory
Coriander	May provide a calming influence if suffering from shock or fear; a gentle stimulant for low energy
Cypress	Helps ease the feeling of loss; creates feelings of security and grounding
Dill	Calms the autonomic nervous system
Elemi	May help with nervous exhaustion, stress and sinusitis; has a balancing and fortifying effect on the psychic centers
Eucalyptus, blue	Stimulating to the hypothalamus; invigorating

Essential Oil	Aromatic Influence
Fir, balsam	Balances emotions; creates a feeling of empowerment; grounding and anchoring
Frankincense	Increases spiritual awareness; helpful for meditation
Geranium	Helps release negative memories and ease tension; spiritually uplifting
Ginger	May influence physical energy, sex, love, money and courage
Grapefruit	Balancing and uplifting to the mind; may help relieve anxiety
Helichrysum	Uplifting to the subconscious; supports feelings of forgiveness and moving forward
Hyssop	Promotes centering and creativity
Jasmine	Uplifting to the emotions; may help increase intuition and wisdom
Lavender	Promotes consciousness, health, love and peace
Lemon	Invigorating and warming; promotes purification and physical energy
Lemongrass	Promotes psychic awareness and purification
Lime	Stimulating and refreshing; May help one overcome exhaustion, listlessness and depression
Mandarin	Refreshing, uplifting and promotes happiness; sedative properties help ease stress and irritability
Manuka	Helps combat stress, calm oversensitive nerves and strengthen the psyche

Essential Oil	Aromatic Influence
Marjoram	Promotes peace and sleep
Melissa (lemon balm)	Calming and uplifting; may help balance the emotions
Mt. Savory	A powerful energizer and motivator; helps revitalize and stimulate the nervous system
Myrrh	Promotes spiritual awareness; uplifting
Neroli (orange blossom)	Encourages confidence, peace, joy, courage and sensuality
Onycha (benzoin)	Warming and soothing to the heart
Orange	Brings peace and happiness to mind and body and joy to the heart
Palmarosa	Uplifting and refreshing; helps reduce stress and tension; increases one's feeling of security
Palo Santo	Clears space of negative energy; calming and soothing to the mind; may help enhance spiritual awareness or meditation
Patchouli	Sedating, calming and relaxing
Peppermint	Purifying and stimulating; studies show it can increase mental accuracy
Petitgrain	Helps to refresh the senses, clear confusion, reduce mental fatigue and support memory
Pine, Scotch	Revitalizes the entire body; soothes mental stress
Rose	Creates a sense of well-being; stimulates and elevates the mind

Essential Oil	Aromatic Influence
Rosemary	Stimulates memory; opens the conscious mind
Rosewood	Creates a feeling of peace and gentleness
Sage	Helps to relieve mental fatigue and strain
Sandalwood	Calms, harmonizes and balances the emotions; supports meditation practice
Spearmint	Helps relieve mental fatigue and strain
Spruce	Helps release emotional blocks; has a grounding effect
Tamanu	Grounding and soothing
Tangerine	Sedating and calming effect to the nervous system
Thyme	Helps supply energy in times of physical weakness or stress; uplifts one's spirit
Valerian	Calming, relaxing, grounding and balancing
Vetiver	Psychologically grounding and calming; supports in the recovery from emotional trauma or shock
White Lotus	Uplifting; creates powerful feeling of well-being
Wintergreen	Elevates, opens and increases sensory awareness
Yarrow	Supports meditation and intuitive energies; balancing
Ylang Ylang	Promotes feelings of self-love, confidence, peace and joy

Why Improve Air Quality?

Sick Building Syndrome Is Real

*A*re you suffering from sick building syndrome, or SBS? If you are, you may experience headaches; irritated eyes, nose or throat; itchiness; dizziness; nausea; fatigue and difficulty concentrating when you are inside your home or office building, but not when you go outside—and others in the building may also have the same experience.

The Environmental Protection Agency recognizes SBS as a real situation in which the conditions inside an office building or home can make people physically ill. Many materials inside a building contribute to SBS, especially in new and recently renovated homes, which are sealed more tightly than old homes and don't allow for better air circulation between the inside of the home and the outdoors. Volatile organic compounds, or VOCs, are released into the air from cleaning agents, carpeting, adhesives, manufactured wood and upholstery, and they can make you feel sick. Mold and bacteria that grow in various places in the home, such as ducts, insulation and carpeting, can also make you sick.

So how can we improve our indoor air quality and reduce toxicity?

Fill Your Home with House Plants

Certain indoor plants make excellent air purifiers. They are able to remove chemicals from the air and turn them into harmless compounds through two natural processes. First, leaves absorb chemicals in the air and transport them to the roots. Second, through the process of transpiration, in which, while excess moisture in the plant is emitted from the leaves, air is pulled down around their roots. Microbes in the roots turn the harmful chemicals into safer compounds. Plants that transpire the most make the best air purifiers.

A NASA study tested fifty plants and ranked the top ten according to their ability to remove three of the most common indoor toxins from the air, their ease of maintenance and their resistance to pests.[6] The three toxins were benzene, formaldehyde and trichloroethylene.

Three Toxins Tested in NASA Study

1. **Benzene:** Found in dyes, detergents, plastics, gasoline and rubber. Benzene is a carcinogen and it causes eye and skin irritation, respiratory problems and kidney and liver damage.

2. **Formaldehyde:** A well-known preservative that is also found in many products, such as insulation, grocery bags, clothing, paper towels and particle board. Formaldehyde causes asthma.

3. **Trichloroethylene:** A compound that is found in adhesives and varnishes. It is a liver carcinogen and damages the central nervous system and respiratory tract.

---∞---

Top Ten Plants in the NASA Study

1. **Areca Palm: The Areca palm tree was** NASA's number one plant for cleaning indoor air. Because of its ability to add moisture to the air, it acts as an excellent humidifier, and it removes toxins such as toluene and xylene from the air.

2. **Lady Palm:** The lady palm tree can also add moisture to the air, and it removes formaldehyde, xylene, toluene and ammonia from the air. This palm tree is very easy to grow, and it is very resistant to pests.

3. **Bamboo Palm:** The bamboo palm removes formaldehyde, xylene and toluene from the air and is easy to care for.

4. **Rubber Plant:** Especially good at removing formaldehyde from the air, the attractive rubber plant is ideal if you lack sunny windows as it needs less sunlight. However, because the leaves are toxic, be cautious with pets and babies.

5. **Dracaena:** The beautiful dracaena excels at taking formaldehyde, benzene, toluene and xylene out of the air.

6. **English Ivy:** This lovely climbing vine can be grown indoors or out. English ivy not only removes benzene, xylene, toluene and formaldehyde from the air but it also removes mold, making it an excellent choice for those with allergies.

7. **Dwarf Date Palm:** The elegant dwarf date palm removes trichloroethylene, formaldehyde, xylene and toluene from the air.

8. **Ficus Alii:** Although the ficus alii is easy to maintain and it removes several toxins from the air, it can irritate people with latex allergies, so those people should wear gloves when handling it.

9. **Boston Fern:** The full, bushy Boston fern is an excellent air purifier, removing formaldehyde, toluene, benzene and xylene, and possibly some mold spores, from the air.

10. **Peace Lily:** Very easy to grow and maintain, the peace lily removes benzene, formaldehyde, trichloroethylene, toluene and xylene from the air.

Having house plants in your home is a great natural way to keep your air cleaner. If you live in a new or recently renovated house, it is especially important to be aware of the toxins around you. When selecting house plants, keep safety in mind and make sure you know which ones may be toxic for pets or children, and keep those out of reach if you do choose one.

The American Society for the Prevention of Cruelty to Animals (ASPCA) has compiled a list of house plants that are non-toxic to both cats and dogs. These are the Areca palm, the bamboo palm, the Boston fern, the dwarf date, the Gerbera daisy, the leopard orchid and the spider plant.[7]

Sound and the Auditory Cortex

---∞---

To enter into the initiation of sound,
of vibration and mindfulness, is
to take a giant step toward
consciously knowing the soul.
~ Don G. Campbell, *The Roar of Silence*

---∞---

*O*ne of the fastest ways to use Air to raise the vibration around you is to introduce different sounds. Sound waves are formed by vibrations in the air created by sounds such as music or words. These sound waves are collected by the outer ear and transmitted to the inner ear, where they move thousands of tiny hairs on a membrane in the inner ear. The movement of these hair cells stimulates nerve cells to send electrical signals to the base of the brain and then up to the auditory cortex, where sound is processed. Different patterns of the electrical signals excite other groups of cells that, for example, can associate the sound of music with feelings, thoughts and past experiences.

Sound transmitted to the inner ear is also broken down according to sound frequencies in an orderly arrangement of low to higher frequencies, much like the way low to high notes appear on a piano keyboard, and different sounds elicit different responses. Some cause a reflexive response, where we may jump, smile or turn our heads. Sounds that are familiar to us are stored in our memories, and we respond to them appropriately. Other sounds can make us feel happy or sad.

We can use certain sound frequencies in our environments to positively influence our well-being. A sound therapy technique called *SomaEnergetics* uses therapeutic tuning forks that vibrate at the exact frequencies required to transform matter to spirit. These are the six frequencies of the solfeggio, which were originally used in ancient Gregorian chants and which can release blockages and facilitate personal transformation.

The 6 Solfeggio Frequencies[8]

Tone	Cycles per Second (Hertz)	Action
UT	396 Hz	Liberating guilt and fear
RE	417 Hz	Undoing situations and facilitating change
MI	528 Hz	Transformation and miracles
FA	639 Hz	Connecting/Relationships
SOL	741 Hz	Expression/Solutions
LA	852 Hz	Awakening Intuition

In the following chapters we explore ways to create positive harmonic resonance using various types of music and musical instruments, as well as the sounds of nature.

Wind Chimes

A garden is a place of peace and beauty, and you can make yours even more peaceful and beautiful with the alluring sound of wind chimes. The sounds created by wind chimes will complete your garden and soothe and enchant you and your visitors. If you don't have a garden, you can choose a wind chime that suits your balcony or front porch.

The Benefits of Wind Chimes

Wind chimes on your porch or patio or in your backyard add a new dimension to your outdoor space. Whether made of metal, wood, glass or other material, they add a gentle ambiance. Hollow wooden tubes provide a particularly appealing and pleasurable sound. The light tinkling sound of wind chimes can provide a soothing, calming and peaceful background noise while you work or play.

There is also evidence that chimes improve our health, and they are used to reduce stress and promote relaxation. The tones they produce are thought to have a healing effect on body and mind, relieving fatigue, comforting a troubled mind and increasing creativity.

The Historical Uses of Wind Chimes

Wind chimes were used in ancient Rome to fend off evil spirits. In all parts of Asia, they were thought to bring good luck, and they are used to increase the flow of chi, or life's energy. Intricate and decorative wind chimes were hung on the corners of temples, palaces and homes in China to ward off birds as well as evil spirits.

Choosing Wind Chimes

Choose a wind chime for the kind of musical tones you would like to have, as well as for its visual appeal. The size and type of material determine the sound. Mini chimes are perfect for small, tinkling sounds, whereas very large chimes will give you a deep, rich, vibrating tone.

Metal, wood, glass, ceramic and bamboo all provide their own unique sound effects. Hard metals such as aluminum and steel produce a sharp tone, whereas soft metals such as copper produce a mellow sound. Glass and ceramic tend to tinkle and ring, and bamboo chimes clack softly and would fit well in a Zen garden designed for quiet reflection. Bamboo is also the most eco-friendly material because it grows fast and replenishes quickly.

You can choose a wind chime that has been tuned to play specific notes, reflecting a particular piece of music or musical era. They can also be tuned in specific keys, such as C or B-flat, and they can have a baritone or a tenor sound.

Most wind chimes are tuned to the *pentatonic scale*, which consists of five notes within one octave. On a keyboard, you can play a pentatonic scale on the black keys, starting with D-sharp for a minor scale, and F-sharp for a major scale.

Solar Wind Chimes

Solar wind chimes can be used indoors, as they do not rely on wind to play; instead, they run on stored energy, which is collected on a solar panel when exposed to sunlight. The sound level is determined by the amount of light they receive.

The Sound of Music

$\mathcal{M}$usic is known to have a profound therapeutic effect on body and soul, and for this reason, music therapy is a growing healthcare field. Music therapists use music to help people improve and maintain their health. Music therapy is beneficial in the rehabilitation of stroke patients and is used in a wide range of institutions, including hospitals, cancer centers, schools and correctional facilities. Music helps people to manage pain, relieve depression and anxiety and ease muscle tension.

The Effect of Music on Our Brain Waves

Studies show that music with a strong beat is stimulating to the brain and causes brain waves to resonate in time with the rhythm, encouraging alertness and concentration, whereas slow beats slow down brain waves and result in a meditative state. This is useful information because it indicates that we can use music to more easily shift the speed of our brain wave activity as needed. One study using light and sound stimulation to treat ADD (attention-deficit disorder) found that rhythmic stimuli sped up brain waves and increased concentration in a way similar to medications such as Ritalin.[9]

The Effect of Music on Our Breathing and Heart Rate

Music can also alter our breathing and heart rates, and therefore activate the relaxation response. For this reason, music therapy can help reduce stress and the damaging effects of chronic stress, which also benefits our overall health.

Studies show that music, such as baroque music, with a pulse between fifty and eighty beats per minute, which falls within the range of our ideal resting heart rate, can help to stabilize our mental, physical and emotional rhythms. Music at this rate decreases blood pressure, encourages relaxation and promotes better concentration and focus. A study using baroque background music in a mathematics classroom found that students enjoyed the class more and found math less challenging with the music.[10]

Other studies have shown that music at sixty beats per minute improves language learning and recall. Because slower music slows the heart beat and breathing rate, lowers blood pressure and relaxes us, it not only enhances our ability to learn but it also creates an environment conducive to creativity.

The Effect of Music on Our Overall Well-Being

Music can help us feel positive and increase optimism and creativity, and because of its beneficial effect on our nervous system, it can reduce the risk of stroke and boost immunity. Music also benefits pregnant women and their babies. A 2007 study showed that pregnant women who listened to lullabies, classical music and sounds of nature, all set at sixty to eighty beats per minute, were less likely than the women who did not listen to these sounds to feel stressed, depressed or anxious.[11]

Not all the benefits of music can be measured, but most of us agree that music can comfort us and help us forget our cares for a while in a safe emotional place. Music can also connects us to a higher power, be it God or Goddess, the Universe, the Source or whatever you wish to call it.

The Healing Power
of Harp Music

———◯◯———

Just as certain selections of music
will nourish your physical body and
your emotional layer, so other
musical works will bring greater
health to your mind.
~ Hal A. Lingerman

———◯◯———

*W*hat do you picture of when you think of a harp? Do you see angels? Many people do, whether they are religious or not. Angels plucking harps has been a common motif in art and religion for centuries. But did you know that the harp has been used for healing since ancient times? The Bible describes David as playing his harp to soothe the king, and harps have been used in royal courts to calm monarchs.

The harp has obvious medical benefits. Anyone who has heard the harp playing knows that it helps them to relax. The range, pitch and rich tone of the harp are such that the music can vibrate through the whole body and create a feeling of comfort and relaxation. Studies show that it can help to reduce blood pressure, increase oxygenation, reduce fatigue, lower the stress hormone cortisol and promote the production of endorphins that can ease emotional and muscle tension.

A study conducted at a cancer center published in *The Harp Therapy Journal* showed that most of the patients in the study found harp music to be calming and relaxing, and a significant number felt energized by it. Many of the nurses and other staff present also reported significant relaxation.[12]

Another study in June 2012 of patients in an intensive care unit found that after ten minutes of harp music, the patients' pain decreased and their blood pressure increased if it was too low and decreased if it was too high, thus demonstrating the harp's stabilizing effect.[13]

So the next time you feel anxious or stressed, listen to some harp music, breathe deeply and enjoy the relaxation.

The Sounds of Nature

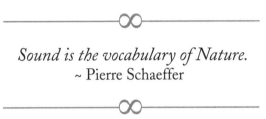

Sound is the vocabulary of Nature.
~ Pierre Schaeffer

*T*here is something wondrous about the sounds of nature. Whether it is the sound of birds singing, frogs croaking, crickets chirping, a gentle breeze in the trees, a rainstorm or waves crashing on a beach, these sounds all affect our well-being. Natural sounds can inspire relaxation or stimulate our creativity.

The sounds of nature help us put our lives into perspective. We are just one small part of a greater, beautiful network. Being outdoors with the natural world around us can make us feel happy and free of stress. Listening to recorded nature sounds indoors can also help us to relax and feel secure. This is a simple, drug-free way to strengthen and balance your body, mind and spirit.

The sounds of nature can be calming or stimulating. Calming sounds, such as birds singing or a gentle rain, can relax us and help us sleep or meditate, whereas more stirring sounds, such as a thunderstorm or howling wolves, can stimulate our brains, making us more alert.

A number of scientific studies demonstrate the healing power of nature's sounds. One study with children from the Alder Hey Hospital in Liverpool, England, showed that using recordings of birdsong and natural sounds such as wind and rain calmed young patients before they received injections and other treatments.[14]

Sound can have a profound effect on our health and well-being, and it is important that we spend time outdoors, where there are only the sounds of nature. When this is not possible, you can benefit from recordings of nature's sounds that you can listen to at home.

Closing Words:
Air and Frequency

---∞---

*Every day we face risks that can affect our
health, but indoor air pollution
is one risk we can do something about.*
~ Dawn James

---∞---

It's easy to overlook it, neglect it or dismiss it; after all, air is invisible, yet is it essential for our existence. The quality of the air around us has a direct impact on our health and well-being. Now you are equipped with the knowledge you need to raise the frequency and quality of the air around you by using air diffusers, filters and purifiers, and transforming the energy of a space using incense, essential oils and sound.

Part II

∞

Light

Not All Light Bulbs
Are Created Equal

————————∞————————

*God created light. Man created
the light bulb, but not all light bulbs
are created equal.*
~ Dawn James

————————∞————————

*I*estimate that the average adult in North America is exposed to
about 3,000 hours of indoor lighting each year, whereas they
get only 400 hours of sunlight annually? In other words, most
people receive 7½ times more unnatural light than natural light.

Our homes, transportation and, for the most part, places of
employment all shield us from the sun. With that in mind, it is
important to understand the various types of indoor lighting that
are available—that you are exposed to each and every day—and
how they affect your vibrational frequency, your mood and, ultimately, your health.

Incandescent Bulbs: The Original!

The incandescent light bulb was born in 1879, thanks to Mr. Thomas Edison. These popular and inexpensive bulbs are the ones we have been using most for lighting in our homes. Powered by electricity that heats a tungsten filament inside the bulb, they provide a warm, steady light that glows. The tungsten filament can be sealed in a vacuum or in an inert gas such as nitrogen or argon. Gas-filled bulbs require more power to get the same amount of light as a vacuum bulb.

The problem with incandescent bulbs is that they require a lot of energy. As a result, governments around the world began to phase out their use and encourage people to buy more energy-efficient alternatives, such as compact fluorescent lamps (CFLs). Brazil and Venezuela began to phase out incandescent bulbs in 2005; the European Union, Switzerland and Australia in 2009; and Argentina, Canada and Russia in 2012. The United States and Malaysia have scheduled phase-outs for 2014.[1]

Before we discard the original indoor light source, let's take a look at some other options, and then you can decide for yourself which would be the better alternative to use in your home. Several organizations defend people's right to have safe lighting. Two groups that come to mind are "Right to Light" and "Spectrum Alliance," both of which have criticized the plans to ban incandescent lamps—primarily because the suggested alternatives have significant adverse effects on human health and the environment.

Fluorescent Bulbs

Fluorescent light bulbs produce more light per watt than incandescent bulbs and, therefore, are more energy efficient. They also come in a variety shapes and sizes. The larger tubular lamps are ideal for areas such as offices, factories, hospitals, schools and retail stores. Compact fluorescent lamps (CFLs) are now available for use in the home. Originally, they produced only cool (white) light; however, the bulbs can now be found in a variety of colors that can change the atmosphere of a room; for example, an amber bulb can replicate the light from a natural gas flame.

Fluorescent bulbs and tubes are filled with a gas containing mercury vapor and argon, xenon, neon or krypton, and the inner surface of the bulbs are coated with a blend of metallic and phosphor salts. When an electric current is passed through the lamps, ultraviolet radiation is produced, which, in turn, bombards the light-emitting phosphorus coating.

What is not widely discussed is that overexposure to fluorescent light can have adverse effects on your health. One reason for this is that the color spectrum of florescent lighting is vastly different from the spectrum of sunlight, which is the most desirable light, and which we will discuss later in this section. Overexposure to fluorescent light has been linked to headaches, fatigue, anxiety, stress, depression and a decrease in sexual function.[2]

Fluorescent lamps also flicker, and even though this pulsing of electricity can be high enough that you aren't aware of it or feel its effects, some people are particularly sensitive to it. The variations in the intensity of the light source disturb their biological systems, and they can suffer from conditions such as headaches, migraines, eye strain and discomfort, nausea, fatigue, sinus problems, blurred vision and irritability.

Fluorescent lamps emit wavelengths of a frequency shorter and faster than those of incandescent bulbs, frequencies that are invisible and reach into the ultraviolet spectrum and are able to penetrate our bodies' cells. This wavelength pattern is much different from full-spectrum light, which is beneficial to us, and can affect us in much the same way as an imbalance in nutrients can have a negative impact on our health. For example, photobiologist John Ott conducted studies that demonstrated the adverse effects of fluorescent light on behavior and academic performance in the classroom, whereby, when fluorescent lighting was replaced with full-spectrum lighting, the students' general attendance, behavior and academic concentration all improved.[3,4]

Compact Fluorescent Lamps (CFLs)

As mentioned above, compact fluorescent lamps (CFLs) are being promoted to replace incandescent bulbs in the home. CFLs have a long life span, and they are among the most energy-efficient bulbs, using 67% less energy than incandescent bulbs. However, they are more expensive, and, like all fluorescent bulbs, they contain mercury, which is not only toxic to the environment when we dispose of them, but there is also evidence that CFLs may be toxic in your home when they are turned on. Despite governments' intentions for citizens to change to energy-efficient lighting sources, CFLs may not be the answer.

CFLs and UV Radiation

In a study published in 2012, researchers at Stony Brook University showed that significant levels of UVC and UVA were emitted from CFLs, and that the emissions appeared to originate from cracks in the phosphor coatings of the lamps, which may be the result of a design flaw in the bulbs. The study revealed that healthy human skin cells responded to the ultraviolet emissions from the CFLs as they did to ultraviolet radiation, whereas incandescent light from the same intensity had no damaging effect on the skin. The researchers concluded, "Despite their large energy savings, consumers should be careful when using compact fluorescent light bulbs. Our research shows that it is best to avoid using them at close distances and that they are safest when placed behind an additional glass cover."[5]

Cancer-Causing Chemicals with CFLs
Tests carried out in Berlin, Germany, led scientists to conclude that several carcinogenic chemicals and toxins are released when "environmentally friendly" CFLs are turned on, in particular phenol, naphthalene and styrene. Peter Braun, who carried out the tests, said, "For such carcinogenic substances it is important they are kept as far away as possible from the human environment."[6]

What If a CFL Bulb Breaks?
Here are some facts from the US Environmental Protection Agency (EPA) to be aware of:[7]

- Never use a vacuum cleaner to clean up mercury. The vacuum will release mercury into the air and increase exposure. The vacuum cleaner will be contaminated and have to be disposed of in a hazardous-waste disposal site.
- Never use a broom to clean up mercury.
- Never wash mercury-contaminated items in a washing machine. Mercury may contaminate the machine and/or pollute sewage.
- If a spill occurs, everyone should be removed from the area, including pets. The impacted area should be sealed off and all ventilation systems in the area should be turned off.
- If a spill occurs, do not walk around as your shoes might be contaminated with mercury. Contaminated clothing can spread mercury around. In case of fire, the entire area will be contaminated with mercury and should be sealed off completely.

- If a spill occurs on a carpet, curtains, upholstery or other like surfaces, these contaminated items should be thrown away in accordance with, and with the assistance of, your local hazardous-waste disposal site.

A Final Word on CFLs

If the risk outweighs the benefits, then stop buying CFLs—that will surely get the attention of and, hopefully, motivate light manufacturers to invent a better light bulb, one that saves energy and poses no health risk to you, your family, your home or the environment.

On personal note, I have consciously and intentionally kept CFLs out of our family home. Why? Firstly, I don't like to jump on the bandwagon when a new product has entered the marketplace until I do my homework. Secondly, I don't want to bring toxins into our home, and, thirdly, I don't want to contribute to the environmental pollution that occurs when CFLs are disposed of.

Perhaps the unified government plan to ban incandescent lights without offering a safe and energy-efficient replacement was premature. What do you think?

Halogen Bulbs

Halogen bulbs are similar to incandescent bulbs. Both contain a tungsten filament that produces light when electricity passes through it. Rather than being surrounded by a vacuum or an inert gas, however, halogen bulbs contain halogen gas. Halogen bulbs are more energy efficient than incandescent bulbs because they consume less energy to produce the same amount of light. They also last longer, and they emit a brilliant white light that is often preferred to the duller light of the incandescent bulb.

There are two main types of halogen bulbs: low voltage and mains voltage.

Low Voltage versus Mains Voltage: What's the Difference?

Low voltage (LV), or dichroic, halogen lamps need to operate through a 240- to 12-volt transformer. As a result, LV lamps are more efficient than the mains voltage (MV) type, which need 220 to 240 volts and do not need a transformer. When lit, the lamps create a similar light; however, there are some differences:

- **Light output:** A 50-watt MV halogen lamp provides about half the light of its LV equivalent. This is fine if you want a soft glow or ambient effect; however, if you have a high ceiling or you want a burst of light, then only an LV lamp will do. In addition, the LV lamps emit a much sharper, whiter light than MV lamps.
- **Light-beam spread:** The MV lamp is available in 25- and 50-degree versions. The LV lamps come in 10-, 24-, 38- and 60-degrees, providing greater flexibility if you wish to light a painting or frame an object.
- **Heat output:** Both lamps emit similar amounts of heat, but in different ways. A 50-watt MV lamp operates at around 85°C, and the heat is mainly projected forward. This may not be desirable if you have a very low ceiling. The LV lamps, on the other hand, direct the heat backwards, so adequate space for air flow is required behind these lamps to avoid overheating.

Xenon Light Bulbs

Low-voltage xenon light bulbs and line-voltage xenon light bulbs are great alternatives to halogen bulbs. These bulbs contain a glass envelope into which xenon gas is introduced, which increases the "rated life" of the bulb.

Low-voltage and line-voltage xenon bulbs last anywhere from 8,000 to 20,000 hours, compared with halogen bulbs, which have an average life of 2,500 to 3,000 hours. Given their long lifespan, xenon lights are great for fixtures that are hard to reach or change.

Unlike halogen bulbs, low-voltage and line-voltage xenon bulbs do not emit ultraviolet rays. This quality makes them ideal for use around sensitive fabrics and artworks. They are also easier to install as they do not produce the high temperatures that halogen lights emit.

Light-Emitting Diode (LED) Bulbs

Unlike the common incandescent light bulb, light-emitting diode (LED) bulbs emit light from a simple semiconductor called a diode rather than from a filament in a vacuum. Incandescent bulbs generate heat, thus wasting energy that could be used instead to generate light, and the filament breaks easily, especially when it gets hot, so the bulbs need to be replaced frequently. LED bulbs are more efficient because they do not generate heat, and they last thousands of hours more than incandescent bulbs—50,000 hours or more, compared with only 750 hours. LED bulbs are also made with plastic, so they are unbreakable. Perhaps the only drawback with LED bulbs is that they have a higher upfront cost.

Light is generated by LED bulbs by the movement of electrons by an electrical current through the light-emitting semiconductor, or diode, which has a positively and a negatively charged component.

Although LED light is often measured in wavelengths, called nanometers, we can also look at them in terms of color temperature. These bulbs are available in different color temperatures, from warm to cool, measured in degrees Kelvin. The color temperature dictates the color of light produced, varying from warm white to cool white (they can produce the same soft white light as an incandescent bulb), and all colors on the color temperature spectrum: red, yellow, green, and blue. For example, if the color temperature a bulb produces is lower, the more yellow your white light will be, and if it is higher, it will cast a cool, bluish hue. The temperature of the light cast in a room affects how the furniture and wall finishes are seen; depending on the temperature, they can appear more blue, pink, yellow, or green.

LED bulbs are also available with colored filters, offering you a rainbow of colored lights to choose from, which brings us to the topic of using colored light bulbs in your home, and how colored lights may affect your heath and behavior.

Colored Light Bulbs

*C*olors can change your mood and behavior. Whereas red can excite and energize you, blue light is often associated with relaxation and meditation. Orange and purple lights are often considered to be spiritual colors. The use of colored lights is an easy and affordable way to enhance the atmosphere and the space around you, whether at home, in your office or in other places where you spend time.

How Are Colored Light Bulbs Made?

Colored light bulbs are made during manufacturing by dipping them into a translucent colored coating. The color coating filters the light to produce only the color the bulb is painted. For example, a green bulb blocks all but the green wavelengths and creates a green light. Fluorescent lights can be colored by placing sheets of transparent colored plastic in front of the fixtures. You can also buy transparent paints for painting light bulbs yourself.

Mixing Colored Bulbs

The primary colors of light are red, green and blue. When a combination of these is mixed, a new color of light is produced. To achieve white light, all three primary colors must be emitted together. Combinations of the primary colors result in the secondary colors of light. These are yellow, magenta and cyan. If a red and green bulb are lit together, yellow light results. When red and blue light are mixed, the result is magenta, and when green and blue are mixed, cyan is produced.

Color and Psychology

We react differently to an object depending on the color of the light shining on it. Restaurants typically use warm-colored bulbs, such as yellow and red, to enhance the atmosphere and to make the food seem more appetizing. Dr. Val Jones, in a research project designed in conjunction with *Architectural Digest*, discovered that blue light can decrease a person's appetite;[8] for this reason, blue and other cool-colored bulbs are not typically used in eating establishments.

Red and other warm-colored lights are considered stimulants, whereas blue is more soothing. According to Pantone, the world-renowned authority on color and adviser on the selection of color for various industries, people are likely to set higher stakes when gambling if there is a red light above them. Blue and other cool colors are typically the choice of medical facilities such as hospitals because of their calming effect.

Color Therapy

Color therapy, also called *chromotherapy*, has been in existence for thousands of years, with documented evidence of "colored halls" and rooms used to treat ailments in ancient Greece, China and Egypt. In 1810, Johann Wolfgang von Goethe published *The Theory of Colours*, describing his systematic study of the physiological effects of color. In 1878, Edwin D. Babbitt published *Principles of Light and Color*, where he described various techniques of healing with color. In the 1930s, an Indian scientist named Dinshah P. Ghadiali published *The Spectro-Chromemetry Encyclopaedia*, which laid the foundation for modern color therapy.

Also in the 1930s, an American doctor, Harry Spitler developed a similar form of color therapy called syntonics. Dr. Spitler discovered that he could generate profound physiological and psychological changes in patients by changing the light that entered their eyes. Several studies have also shown that different colors can affect our blood pressure, pulse and respiration rates, biorhythms and brain activity.[9] Because of this, various ailments are now being treated using colors.

Colors put out vibrational energy, which our bodies absorb, and we can obtain all the energy we need to maintain healthy bodies, minds and souls through color. Our mental health and efficiency in life depend on normal color balance; in other words, if we are out of balance, it is possible to strengthen our energy centers using color. Our energy centers are connected to all our organs, body systems and functions. This is an example of how the mind-body network operates.

Light energy consists of the seven colors: red, orange, yellow, green, blue, indigo and violet. Each color is connected to certain areas of our bodies and affects us emotionally, physically and mentally. If you learn how each color can influence you, you can use color for an extra boost of energy when you need it.

Some color therapists believe that the energy vibrations of color have healing properties; therefore, exposure to a color and its vibrations can be used to assist our bodies' natural healing and recuperative powers to achieve and maintain health and well-being.

Physical healing is encouraged by directing colored light toward diseased areas of the body or toward the eyes. In color therapy, each color corresponds to one of the seven chakras (energy centers in the body), which, in turn, can influence a specific gland, organ or tissue.

Effects of Colored Light on Our Physical and Energy Bodies*

Color	Psychological Effect	Physical Effect
Red	Vitality, courage, self-confidence	• Stimulates the root chakra at the base of the spine, causing the adrenal glands to release adrenalin, which results in greater strength • Causes hemoglobin to multiply, increasing energy and raising body temperature • Excellent for anemia and blood-related conditions • Relieves clogs • Releases stiffness • Stimulates the liver • Builds energy • Increases circulation
Orange	Happiness, confidence, resourcefulness	• A blend of red and yellow, orange combines physical energy with mental wisdom • Treats conditions such as inflammation of the kidneys, gallstones, prolapses, menstrual cramps, epilepsy, wet cough and all sinus conditions • Helpful in dealing with excess sexual expression

Color	Psychological Effect	Physical Effect
Yellow	Wisdom, clarity, self-esteem	• Stimulates the solar plexus, or psychic center • Can be used for psychic burnout or other psychic-related conditions or ailments • Helpful for conditions of the stomach, liver and intestines • Helps the pores of the skin and aids the healing of scar tissue itself • Typically used to treat constipation, gas, liver troubles, diabetes, eczema and other skin troubles, leprosy and nervous exhaustion • Has an alkalizing effect, which strengthens the nerves
Green	Balance, love, self-control	• Corresponds with the heart center, healing heart troubles and decreasing and stabilizing blood pressure • Heals ulcers, cancer, headaches, nervous disorders and influenza • Acts as a general tonic • Provides a feeling of renewal, peace and harmony • Connects us to unconditional love and is used for balancing our whole beings

Color	Psychological Effect	Physical Effect
Blue	Knowledge, health, decisiveness	• Stimulates the throat chakra; can be used for any type of ailment associated with speech, communication or the throat • Stops bleeding of the lungs • Decreases fevers • Cures sore throats • Provides relief to most inflammations of the skin and gums, and can be used with infants for pain while teething • Used for goiter, measles, chickenpox, cuts, bruises and burns • Bring great calm and peace to the mind that is worried, excited or in a constant nervous state
Indigo	Intuition, mysticism, understanding	• Stimulates the brow chakra (third eye) and controls the pineal gland, which relates to ailments of the eyes and ears • Particularly beneficial in treating cataracts, glaucoma and various eye problems • Purifies the blood and the mind • Can be used as a remedy for ear and nose complaints, diseases of the lungs, asthma, infantile convulsions and mental complaints

Color	Psychological Effect	Physical Effect
Violet	Beauty, creativity, inspiration	• The color of the divine spirit and links with the crown chakra • Excellent remedy for neurosis, diseases of the scalp, sciatica, tumors, rheumatism, cerebral-spinal meningitis, concussion, cramps and epilepsy • Leonardo da Vinci proclaimed that you can increase the power of meditation ten-fold by meditating under the gentle rays of violet
White	Balance, harmony	• This is the color of the awakened Spirit • Directing white into the aura helps stimulate the person's own divine nature into healing the self

*Adapted from Altered States, "Color Therapy," (n.d.)
http://altered-states.net/barry/newsletter220.

Colored-Light Therapy Guidelines

The general rule of thumb is to place the affected area approximately ten to twelve inches from your light source if you are indoors. Ideally, you should apply the therapy twice a day; however, listen to your body and recognize when the colored-light therapy has served its purpose and is no longer needed.

Duration of Colored-Light Therapy

- **Red:** Five to ten minutes (never apply red to the head area)
- **Orange:** Five to fifteen minutes (fifteen minutes only for sinus problems)
- **Yellow:** Fifteen minutes
- **Green:** Ten to twenty-five minutes (this is the only color that can be applied for such a length of time)
- **Blue:** Five to fifteen minutes (never overexpose the area of the head)
- **Indigo:** Ten minutes (for eye therapy, one to five minutes is usually sufficient)
- **Violet:** Five to twenty-five minutes (use for twenty-five minutes only if treating sciatica, and expose only the back or sides of your body, not the front)

Note: Any color that is applied to a specific area must be localized; this is very important. Green, yellow and blue may be general.

Candles

$\mathcal{W}$hether you are preparing to slip into a bathtub or setting a dining table, lit candles can transform a room within seconds. Candles can also be used for aromatherapy, to create a pleasant scent, to add romantic ambiance, for decorative purposes or for ceremonies and magick.

There are several types of candles that are ideal for certain uses, but the candles of the highest vibration that I have found and use regularly in my home are candles made from 100% pure cappings beeswax. But more about that in a moment.

Types of Candles

First, let's take a look at a few popular kinds of candles and how they can be used.

- **Pillar candles:** These candles are cylinder-shaped wide candles. The wax is rigid enough to burn free-standing (on a heat-resistant base). They are available scented as well as unscented. They are popular for creating relaxing and romantic ambience in a room.
- **Taper candles:** These candles are designed to be used in candlesticks or in standard decorative wall sconces. They burn gently for several hours.

- **Votive candles:** These are smaller candles, typically used in lanterns or ornamental candle holders. They burn rapidly and need to be placed in a secure container.
- **Floating candles:** Placing a candle to float on water creates a beautiful visual effect as the flame is reflected in the water; it is also one of the safest and secure ways to burn a candle.
- **Jar candles:** These are scented candles, usually in a base of paraffin or soy wax.

Types of Wax

It's important to understand the type of wax that a candle is made of as it acts as the fuel for the candle, and its output has an immediate effect on your indoor air quality.

- **Paraffin wax:** Paraffin is a by-product of crude oil and one of the most dominant types of wax used for candles. It comes in a variety of grades distinguished by their melting points. For instance, a paraffin wax labeled as having a low melting point (i.e., it melts at or below 130°F) is a very soft wax suitable for making candles in containers. A medium melting point paraffin changes from solid to liquid between 130°F and 145°F. This wax can be used for making poured candles. A higher melting point of 145°F to 150°F is a harder wax, suitable for carving or molding. A general rule of thumb is that the higher the melting point, the harder the wax and longer the burn time.
- **Soy wax:** Soy wax is a cleaner-burning wax compared with paraffin. It has gained popularity as an inexpensive natural wax. The melting point of soy wax ranges from 120°F to 180°F, depending on the blend. Soy wax tends to hold scents better and longer than paraffin wax.

- **Beeswax:** Beeswax is the oldest and purest form of wax in the world. Some of the advantages of beeswax are that it is virtually smokeless, it is naturally sweet scented, it burns slowly and it has a high melting point.

Buyer Beware: Is It Really Beeswax?

A word of caution about candles labeled "beeswax": By law, a candle containing as little as 10% beeswax can be legally labeled and sold as a beeswax candle. If a beeswax candle contains other ingredients, such as paraffin, then it will not produce negative ions, which have a positive effect, as explained below.

My Favorite Beeswax: Cappings Beeswax

Once bees have built their honeycombs and filled them with honey that they convert from pollen and nectar, the field bees feed nectar to a few of the younger ones because they are especially efficient at using it to produce wax, which appears as small white flakes on their abdomens. These young bees, only ten to sixteen days old, masticate the wax flakes and apply it as cappings to the filled honeycombs, signaling to the beekeepers that the honey is ready for harvesting.

To obtain the honey, the cappings have to be gently removed from the honeycomb using a special knife and being careful to brush off the bees that are still clinging to it. It takes ten pounds of honey to produce about one pound of cappings beeswax that can be used to make candles.

Some Benefits of Cappings Beeswax Candles

- They emit negative ions when they burn, which counteract the positively charged toxins, dust, pollen, odors and molds, and even viruses that float in the air around us, whether at home or in the office.

- They burn cleanly and are non-toxic and non-carcinogenic, and do not cause allergies. You can think of them as indoor air purifiers that also produce a warm glow in your room.
- They cost only pennies an hour to burn, lasting longer than other candles, including vegetable or soy waxes.
- They look pretty, with their bright, golden halo, and they exude a delicious aroma of honey essence.
- They are proven to be of benefit to those who have allergies and environmental sensitivities, and they help relax you when you are feeling stressed.

Sun-Loving

———————∞———————

Although we are millions of miles apart,
I still feel your warm caress.
The song birds announce your arrival
in the east,
While petals close to bid you farewell
in the west.
You shower me with blessings
from sunrise to sunset—
Light, fuel, food, heat, energy and more,
yet
You ask for nothing in return—
The unconditional love of the Sun.
~ Dawn James

———————∞———————

For thousands of years, ancient cultures around the world revered the sun as a life-giving, healing power. But it wasn't until 1877 that scientific study of the sun was taken seriously, when Downes and Blunt discovered the dramatic ability of the sunlight to destroy bacteria. Fast forward another twenty-six years, and in 1903, Neils

Finsen won the Nobel Prize for successfully treating skin tuberculosis with ultraviolet rays. Since then, studies have continued to reveal the numerous benefits that sunlight has on the human body.

What is it about a sunny day that makes everyone smile more and walk with a little more pep in their step? Perhaps it's because the sun provides an abundance of benefits to our psychological, emotional and physical well-being.

Here are just a few benefits from the sun:

- Lowers elevated blood pressure
- Decreases cholesterol in the blood stream
- Increases disease-fighting white blood cells
- Lowers blood sugar
- Increases energy, endurance and muscular strength
- Increases the oxygen-carrying capacity of blood
- Increases the number of red blood cells (beneficial for anemia)
- Increases cardiac output (a slower, more efficient and rested heart pumps more blood)
- Increases sex hormones
- Increases resistance to skin infections
- When sunlight enters the eyes, it helps in the regulation of melatonin production

Sunlight's Effect on Indoor Air Quality

We don't often think that sunlight can affect air quality. Did you know that when sunlight passes through the atmosphere, it electrically charges some of the air molecules? Some air molecules will become positive ions and some will become negative ions.

Certain equipment, such as heaters and air conditioners, typically remove negative ions and usually add positive ones into the air. The problem with this is that as positive ions increase and the negative ions decrease, some people begin to feel adverse effects, such as headaches, dry throat, sinus problems, hoarseness, fatigue and dizziness. Negatively charged air on the other hand produces feelings of contentment and well-being.

Negatively Charged Air and Cancer

One fascinating study I found in my research was in the journal *Cancer Research*, which reported that negatively charged air can inhibit cancer growth. A test was done on a group of rats with cancer. Half the rats were allowed to breathe negatively charged air, while the other half breathed regular indoor air. After one month, the cancer grew to twice the size in the rats breathing indoor air versus those that were breathing negatively charged air. But the most fascinating discovery was when they introduced the supplement calcium lactate to the rats' diet in addition to their breathing negatively charged air, and the cancer growth completely stopped![10] Now, this study is over fifty years old, which begs the question: Why do companies continue to spend billions on cancer research?

Sunbathing and Free Vitamins

We need vitamin D to absorb calcium, and thus protect our bones and teeth, and it is also protective against cancer and many other diseases. The best source of vitamin D, as is commonly known, is from sunlight on our skin; therefore, sunbathing is a natural way to get this vitamin. Approximately half an hour of summer sun without sunscreen two or three times a week, wearing just shorts or a swimsuit, gives you an adequate amount of vitamin D. However, this depends on the hemisphere in which you live, and you may need to get more sunshine and even take vitamin D supplements the further away you are from the equator, especially in the wintertime. If you can only bare your shoulders, arms and legs, then half an hour of summer sunshine four or five times a week will probably be enough. It is important not to use sunscreen because it blocks the UVB rays that are needed for vitamin D synthesis. It can be helpful, especially in winter, to get your doctor to order a lab test for your vitamin D levels to determine if you need to take a supplement.

Seasonal Affective Disorder (SAD)

Lack of sunshine can also lead to a mood disorder called seasonal affective disorder (SAD), also known as the "winter blues," which affects millions of people worldwide. SAD is associated with episodes of depression, which occur during seasons of less sunlight; that is, in the winter months. Typical symptoms of SAD include the following:

- Lack of energy
- Depression
- Fatigue
- Cravings for sweets and weight gain

Melatonin is a hormone that is secreted by the pineal gland in the brain and helps to promote sleep at night. One method that many people suffering from SAD respond positively to is the use of "phototherapy," or "full-spectrum light therapy." This light therapy replicates the full spectrum of sunlight, which can lead to the proper synthesis of melatonin and re-established circadian rhythms, which helps people to sleep at night and has an antidepressive effect. In essence, a few hours of full-spectrum light in the fall and winter days can fool the body into thinking it is spring or summer.

Sun Light, Cholesterol and Heart Disease

Not only does direct sunlight provide our bodies with vitamin D, but it can also convert cholesterol in our skin into vitamin D. This is because the molecular structure in cholesterol (specifically, 7-dehydro-cholesterol) is similar to the molecular structure of vitamin D. Our skin contains a significant supply of cholesterol, and it moves back and forth between our skin and our blood stream. Medical studies have shown that when our skin is exposed to sunlight (for as little as 20 minutes a day), there is a reduction in cholesterol in the skin, and subsequently the overall cholesterol metabolism in the body is increased, which, ultimately, lowers blood cholesterol.[11] The key is to consistently get a daily dose of unfiltered sunshine.

Sun-Gazing: Myth and Facts

Sun-gazing is an ancient practice that has existed as long as humans have occupied planet Earth. In 1992, Hira Ratan Manek (aka HRM), from India, shared the benefits of sun-gazing with the Western world. Between 1962 and 1992, HRM studied the ancient practice of sun-gazing and rediscovered the fact that we can activate parts of the brain that are currently dormant. He also discovered that we can awaken inherent gifts and powers within ourselves to higher levels, in essence, raise our frequencies and increase our inherent creative abilities and willpower.

Sun energy is a power source that activates the brain through the eyes—the eyes act as a doorway. Following are some myths and facts about sun-gazing:

> **Myth:** Staring at the sun can damage our eyes.
> **Fact:** Safe sun-gazing does not harm the eyes; it makes them healthier. Sun-gazing practices of the Mayans, Egyptian and

Incan priests and priestesses, Native Americans and monks serve to remind us of this truth.

Fact: The safest time to sun-gaze is the first hour of sunrise, or within one hour of sunset.

Fact: The absorption of sun energy becomes your new food! It substantially decreases your need for food because it makes you feel less hungry. Remember, food is not necessary for the body to function; only energy is!

Fact: In 1922, the medical faculty at Imperial College in London, England, decreed that solar rays were the ideal food for humans.

Some Benefits of Sun-Gazing

1. It activates areas of the brain that are currently dormant via the hypothalamus.
2. We receive vitamins A and D during the first hour of sunrise.
3. It brings balance to the mind and encourages a positive attitude.
4. It increases self-confidence.
5. It helps you develop your inherent powers.
6. Problem solving becomes easier with increased brain function and comprehension and heightened intuition.
7. It strengthens the hypothalamus, which acts as a critical link between the nervous and the endocrine systems via the pituitary gland.
8. It decreases your appetite for traditional food, while increasing energy levels.

Guidelines for Sun-Gazing Practice[12]

Sun-gazing is a one-time practice that is typically done over a nine-month period in three phases, depending on the climate in your area. As a precaution, have your eyes examined before you consider sun-gazing. If you cheeks flush while doing this, stop. Listen to your body! Practice the following in bare feet on bare earth, not on grass.

Phase I (0–3 months)
Day one: Look at the sun for a maximum of ten seconds during the first hour of sunrise.
Day two: Look at the rising sun for twenty seconds, then add ten seconds each day thereafter. Eye stillness and steadiness are not required; do not wear contact lenses or glasses while sun-gazing.

Phase II (3–6 months)
As you reach thirty minutes duration of looking at the sun, you will slowly be liberated from certain physical imbalances, illness and diseases as all the colors of the sun will have reached your brain through your eyes. Your brain redirects the flow of color prana appropriately throughout your body, where it is needed. Your vital organs are dependent on certain sun color prana; for example, the kidneys = red, the heart = yellow, the liver = green. This color prana reaches the organs and corrects deficiencies and imbalances and weaknesses—this is how color therapy works! At this phase your appetite for food decreases.

Phase III (6–9 months)
The absorption of sun energy becomes your new food! Need for food intake decreases substantially as you feel less hungry. Remember, food is not necessary for the body to function—only

energy is! By eight months, the concept of hunger will no longer exist. All mechanisms associated with hunger, such as aroma, cravings and hunger pangs will also disappear. At nine months, you will have reached a total of forty-four minutes of sun-gazing, which is the maximum daily intake recommended by solar science to ensure proper eye care.

While the sun provides many benefits for our physical well-being, the moon similarly provides many benefits for our spiritual well-being.

By the Light of the Silvery Moon

------------∞------------

The moon is at her full, and riding high,
Floods the calm fields with light.
The airs that hover in the summer sky
are all asleep tonight.
~ William C. Bryant

------------∞------------

*W*hen we learn how to work in harmony with the phases of the moon, we can enhance our creative abilities and manifest wonderful things. Following the lunar cycle can enhance your existing spiritual practice or simply deepen your connection to the Earth and the cosmos.

There are eight lunar phases of the moon: new moon, crescent moon, first-quarter moon, gibbous moon, full moon, disseminating moon, last-quarter moon and balsamic moon. Each moon phase represents an aspect of the planting cycle (from seed to growth to bearing fruit), and each phase also corresponds to one of the eight seasons (we have four major seasons and four minor seasons).

Here are some guidelines for working in harmony with each moon phase:

The New Moon: The lunar phases begin at the new moon. It marks new beginnings. It's a time for gathering our thoughts, planning our intentions and preparing to announce them at the crescent moon.

The Crescent Moon: The crescent moon is the time to announce our intentions, hopes and wishes for the lunar month. It is good to write them down either on a piece of paper or in your journal. Journaling can be a valuable tool for spiritual practice. Remember to say your intentions out loud.

The First-Quarter Moon: The first-quarter moon appears approximately a week after the new moon. When the moon is in the first quarter, it is in a building mode. The first quarter moon asks you to make an action plan to enhance your manifesting power. Taking action toward your goals adds energy to the actualization of your list of intentions.

The Gibbous Moon: The gibbous moon is three days before the full moon. It is now time to increase your ability to receive what you have intended by visualizing fulfillment. Review your list, close your eyes, go into meditation and visualize fulfillment.

The Full Moon: This is the time that your contract is signed, sealed and stamped. Note that it is not necessarily delivered. It may be, but if it is not, do not be discouraged. You still have two weeks to devote to this intentional period. Know that the intention is "out there" in the Universe, and time is gathering all the circumstances and other people's intentions and time frames together to make this happen the best way for all concerned. Take time to acknowledge what was manifested from your list.

The Disseminating Moon: This is the time to be grateful, talk about your intentions to those who wish you well, share your hopes and wishes and ask those who love you to keep good thoughts for you. Tell yourself enthusiastically that your intentions are "already done." You are now entering the releasing phase.

The Last-Quarter Moon: When the moon is in the last quarter, it is a time designated for releasing habitual patterns. This is an optimal time to let go, release and forgive anything and everyone that you may feel has hurt you in any way. It is a time of emptying self and allowing the Universe, or whomever you trust in your faith, to take over. It is out of your hands and no more action is needed from you.

The Balsamic Moon: The balsamic moon is seventy-two hours before the new moon. This is a time to enter the void, to transmute, to empty out in order to create space and prepare for the new cycle. If you have written a list, now is the time to burn it. This symbolizes your releasing process going into emptiness. This is the time to surrender, rest and recuperate. No more thinking, planning, getting in your own way or action of any kind—just peace. As the moon slowly enters "the dark of the moon," we allow our visual minds to become dark, too, and still.

As the moon moves through its phases, it acts as a subtle visual reminder to us that life is constantly changing. The key is to live in the natural giving and receiving of life, with reverence and acceptance, allowing and not resisting change—not as a passive bystander, but as an active participant.

Closing Words:
Light and Frequency

------------∞------------

*All forms of matter are really
light waves in motion.*
~ Albert Einstein

------------∞------------

*H*uman beings, plants and animals have a biological clock and it tells each human, plant and animal when to eat, sleep and wake up. In humans, the circadian rhythm is regulated by several genes, hormones and the environment. Light is the major environmental modulator of the human circadian rhythm through its effect on the hormone melatonin. Melatonin has many functions in humans, including the regulation of the sleep-wake cycle, immune system and metabolism, and it has anti-cancer properties, such as scavenging free radicals and activating antioxidative pathways.[13]

Exposure to artificial light at times of normal darkness, "light at night," suppresses melatonin and disrupts the natural circadian rhythm. This can lead to sleep disturbance and sleep disorders. These disruptions not only lower our personal frequencies, resulting in fatigue, but they also result in immunosuppression that has been shown to be linked to increased cancers.

It is known that white artificial light, at certain wavelengths, suppresses the production of melatonin in the brain's pineal gland. In addition, suppressing the production of melatonin, which is responsible, among other things, for the regulation of our biological clocks, causes behavior disruptions and health problems.

And to this end, I hope you will agree that it's critically important that we seek new ways of increasing the ratio of natural light to artificial light in our lives.

Part III

∞

Water

The Vibration of Water

*W*hether it's falling from the sky, pumped from the ground, delivered by the truckload or delivered from a water-treatment facility to your tap, all water carries a vibrational frequency. A number of factors determine this frequency, including the conditions the water goes through before arriving at your tap and the water source (for example, a lake, the local the aquifer your well connects to or the water-treatment process).

The vibration of water from a residential well is vastly different from the vibration of water processed in a water-treatment plant. One reason for this is that the water from a well has a short life span—it's consumed only once (from a well to the taps to a septic system in the ground)—whereas "city water," or water recycled through a treatment plant, typically has several life spans, which causes frequencies to accumulate each time the water is treated.

If you live in the city, have you ever thought about how many times the water coming out of your tap has been recycled? Does recycling remove all impurities? Is there something more you need to do to ensure your water is clean and safe? In recent years, there have been numerous news reports and scientific studies about the presence of estrogenic compounds (natural estrogens and synthetic chemicals that mimic natural estrogen) in waterways and drinking water and their potential harm to human health and aquatic life.[1] At first, these claims focused only on birth control hormones found in "treated" water, but if you dig a little deeper, you will uncover a number of sources of estrogenic compounds, also referred to as endocrine-disrupting compounds in our water, including synthetic estrogens in crop fertilizer (e.g., atrazine); synthetic and natural estrogens from livestock, including dairy cows, which can be fed hormones to increase milk production; and an unknown number of industrial chemicals, such as the plastic additive bisphenol-A (BPA). In addition to estrogenic compounds, traces of pharmaceuticals also appear in drinking water and sewage water, including anti-depressants, painkillers and antibiotics. What are the long-term effects of these substances in our drinking water? Ironically, in the United States, for example, most pharmaceutical chemicals are not regulated by the Environmental Protection Agency. In other words, no governmental agency in that country regularly tests for pharmaceuticals in public drinking water.

Now I don't want to leave you with a false impression that country dwellers have the safest water either. The more rural the setting, the higher the chances are of livestock and industrial farming operations existing in the area. Country dwellers need to know about their local farming practices to understand which chemicals or compounds may potentially be seeping into the aquifers that supply water to residential wells. For instance, many plant fertilizer and soil products contain treated bio-sludge, which also contains, of course, human waste. These bio-sludge products undoubtedly contain traces of pharmaceuticals as well. Take this example one step further and ask yourself what happens to crops grown with soil treated with bio-sludge.

It's important to be aware of these issues so you can take commonsense steps to reduce your exposure to unwanted and dangerous substances. Clean water is essential for good health, and in the section on page 83, "How to Improve Water Quality at Home," I outline several things you can do to improve the quality and raise the vibration of the water in your home.

Problems with Chlorinated and Fluoridated Water

*W*ater is essential for our bodies to function properly. Unfortunately, most people in North America drink and use unfiltered, chemically treated water (i.e., municipal or city water) in their homes and work spaces. Two of the predominant chemicals found in city water in North America are chlorine and fluoride—both of which pose significant health risks.

Dangers of Drinking Chlorinated Water

Chlorine belongs to a group of five chemicals in the periodic table called halogens. The other four are fluorine, bromine, iodine and astatine. Of these, chlorine, bromine and iodine are often used as disinfectants.

Chlorine gas is obtained from the salt sodium chloride through the process of electrolysis, and then, under pressure, the gas is liquefied. This liquid is a strong oxidizing agent that can kill pathogens that are commonly found in water and that can cause disease.

Before the chlorination of our drinking water, thousands of people died every year of diseases such as cholera, typhoid fever and dysentery, and then chlorine was introduced as its use was thought to be safe. However, it is now known that chlorine reacts with organic compounds that are naturally found in water and produce what are known as disinfection by-products (DBPs).

One of the common DBPs are trihalomethanes, which are potentially carcinogenic, causing liver and kidney cancer, as well as heart disease.

A study published in 1992 in the *American Journal of Public Health* concluded that there was a positive association between the consumption of chlorination by-products in drinking water and bladder and rectal cancer in humans. The researchers noted that about 9% of all bladder cancer and 18% of all rectal cancer cases could be linked to the long-term use of chlorination by-products.[2]

Another study concluded as early as 1987 that "long-term drinking of chlorinated water appears to increase a person's risk of developing bladder cancer by as much as 80%."[3]

Dangers of Drinking Fluoridated Water

Water fluoridation is a process of adding fluoride to the public water supply for the purpose of reducing tooth decay. The rationale is that fluoride reduces the rate of demineralization of tooth enamel and increases mineralization to prevent cavities. However, did you know that water is fluoridated for only about 5% of the world's population, with more than 50% of these people living in North America, whereas Europe has banned or stopped fluoridation because of environmental, health, legal or ethical concerns?

Studies show that fluoridated water negatively affects IQ in children, age of puberty, fertility rates, mental health, thyroid function and more, even dental health! Following are some of the studies that illustrate the dangers of fluoridated water:

- Many studies that have been done around the world have shown that fluoride affects intelligence in children. Two Chinese studies concluded that even low levels of fluoride exposure in drinking water have negative effects on children's intelligence, as well as on the development of dental flurosis (mottling of the teeth).[4,5]

- A research study done for a doctoral dissertation at the School of Biological Sciences, University of Surrey, England, on the effect of fluoride, which accumulates in the pineal gland of humans, found in an animal study that fluoride in the pineal gland accelerated the onset of puberty.[6]

- An epidemiological study in several regions of the United States in 1994 showed that decreasing fertility rates in women was associated with increasing fluoride levels in the drinking water.[7]

- In a study that compared serum testosterone levels in (1) men afflicted with skeletal fluorosis who lived in fluorosis-endemic areas, (2) men who were healthy but drank the same water and (3) healthy men who did not live in fluorosis-endemic areas, the researchers concluded that fluoride toxicity may cause adverse effects in the reproductive system of males living in fluorosis-endemic areas.[8]

- In their review of the Environmental Protection Agency's standards for fluoride in drinking water, the National Research Council noted that because fluoride increases the production of free radicals in the brain, it is possible that it can increase the risk of developing Alzheimer's disease because an increase in free radicals is thought to be associated with this disease.[9]

- One of the earliest fluoridation trials in the United States reported in the *Journal of the American Dental Association* in 1956 that girls living in a community with fluoridated water reached menarche five months earlier than girls in a nearby community that did not fluoridate its water, and that children in the fluoridated community had cortical bone defects and irregular mineralization of the thigh bone.[10]

- A study published in Moscow in 1985 concluded that pro-longed consumption by healthy people of drinking water with raised fluorine caused elevated thyroid stimulating hormone (TSH), decreased T3 concentration and more intense absorption of radioactive iodine by the thyroid as compared with healthy people with normal fluorine content in their drinking water.[11]

Following are two interesting facts about the use of fluoride and its impact on thyroid function:

1. In the mid-twentieth century, fluoride was prescribed by a number of European doctors to reduce the activity of the thyroid gland in patients with hyperthyroidism (overactive thyroid).[12]

2. In a report in 1991, the US Department of Health and Human Services estimated that total fluoride exposure in communities whose water is fluoridated ranges from 1.6 to 6.6 mg per day from all sources.[13] Could this be the link to the increasing problem of hypothyroidism (under-active thyroid) in the United States and other fluoridated countries?

How to Improve Water Quality at Home

*I*am a firm believer that to be forewarned is to be forearmed. Now that you know what may be lurking in your tap water, it's critical to explore some options you can take to reduce toxicity in your home, in your body and in your life.

Whether you are dependent on water from a water-treatment facility or from your local well, there are several compelling reasons why you need raise the frequency of your water—through filtering, purifying and/or conditioning—to make it cleaner and safer.

The first step is to take a sample of your water to your local health department for analysis to learn to what extent your water may need additional "conditioning." The next step is to determine how large a project you want to undertake. Do you want to condition all the water coming into your home, or will one tap in the kitchen work if it's the only tap water you'll use for drinking?

Following are some solutions for your whole house and for a single tap in your home. Each option addresses a particular water-quality issue, depending on your needs.

Whole-House Options

Sediment Filter

Whole-house sediment filters designed to remove dirt, sediment, rust particles and other particulates from city and well water improve water clarity, taste and odor, and they protect your home's plumbing system (including other water-treatment equipment) from damage and premature clogging due to sediment buildup.

You can use simple plumbing fittings from your local hardware store to connect to your pipes if you have ¾-inch or ½-inch pipes. But check your water flow rate before obtaining a sediment filter as they are two types: "regular rate" and "high-flow rate." In addition to sediment filters, if your water tests show high iron, manganese or hydrogen sulfide content, consider getting a whole-house iron/hydrogen sulfide reduction filter instead. These filters should be installed on the main cold-water line after installing the pressure tank (for country dwellers) or water meter (for city dwellers).

Ultraviolet (UV) Light

Ultraviolet water purifiers are simple to use and they effectively destroy microorganisms in water. These systems are also often combined with a water filter to trap even more contaminants because UV light is capable of penetrating the cell walls of bacteria, viruses, molds, yeasts and parasites such as Cryptosporidium and Giardia, disrupting their ability to replicate, but they do not remove chlorine, volatile organic compounds (VOCs), heavy metals and other chemicals.

To maintain the effectiveness of a UV system, it is recommended that you replace the UV light annually.

Some benefits of UV systems include the following:

- They do not introduce any chemicals or by-products into your water.
- They do not alter the taste, pH or other properties of your water.
- They are not harmful to your plumbing and septic systems.
- They are easy and cost-effective to install and maintain without any special training.

Water Softeners

If the water in your home is "hard," that is, it is high in dissolved minerals such as calcium and magnesium, then you should consider using a water softener. Hard water results in scale buildup on your fixtures, in your kettle and in water pipes, which restricts water flow. It can also coat heating elements, leading to power loss and increased electricity consumption. Soap doesn't lather well in hard water, resulting in a residue on your hair and skin. A water-softening system reduces the water hardness by replacing the minerals with sodium or potassium salts.

Water softeners consist of a brine tank that is filled with negatively charged polystyrene "beads," also known as resin, that are bonded to positively charged sodium or potassium ions. As the hard water flows through the tank, the sodium ions change places with the calcium and magnesium ions because they have a stronger positive charge than the sodium ions. Once the calcium and magnesium have replaced all the sodium in the beads, the water softener begins a "regeneration cycle," during which the beads are soaked in a strong solution of water and sodium chloride or potassium chloride, which recharges the beads, and the water softener flushes the remaining brine and the hard minerals through a drainpipe.

A water softener is the most expensive investment of the whole-house options with regard to the cost of the equipment and the ongoing purchase of salt to fill the brine tank. If you choose this option, invest wisely by selecting the right-sized unit based on the number of people in your household and the daily average water consumption.

Single-Tap Options

Reverse Osmosis plus Remineralization

Reverse osmosis (RO) is a water-purification process whereby water moves through specialized membranes to remove foreign contaminants, solid substances, large molecules and minerals, including lead.

The downside of RO as a method for purifying drinking water is that healthful, naturally occurring minerals and trace minerals in water are also removed. These minerals not only provide a good taste to water, but they also serve a vital function in the body.

When stripped of trace minerals, water can actually be unhealthful for the body. Yes, drinking "pure" water can be unhealthful because the fewer minerals we consume, the greater our risk for osteoporosis, osteoarthritis, hypothyroidism, coronary artery disease, high blood pressure and a long list of degenerative diseases generally associated with premature aging.

The ideal water for the human body should be alkaline, and this requires the presence of minerals such as calcium and magnesium; for this reason, water that has been purified by reverse osmosis needs to be remineralized.

One of the most effective solutions I have found and personally use comes from MineralPRO Drinking Water Systems.

MineralPRO manufactures whole-house and drinking-water systems and bottleless water coolers that purify the water and add the natural minerals back into the water (remineralization). They use naturally occurring organic minerals that are common to natural water sources, including calcium, magnesium, potassium and sodium, which naturally correct the pH only enough to reach a healthful alkaline state.

Remineralizing your drinking water has significant health benefits, including the following:

- **Calcium** is required for strong teeth and bones as well as nerve transmission, blood coagulation and muscle contraction. The optimum daily intake of calcium for adults is 1,000 to 1,500 mg.

- **Magnesium** plays many essential roles in the body, including the control of muscle contraction, protein metabolism, blood coagulation and energy production, and it is required for the absorption and metabolism of calcium. Lack of sufficient magnesium can lead to high blood pressure and osteoporosis. The optimum daily intake of magnesium for adults is 400 to 600 mg.

- **Potassium** is critical for proper electrolyte and acid-alkaline balance. This mineral helps reduce the risk of high blood pressure and stroke, and it plays a role in reducing feelings of anxiety, irritability and stress. Lack of potassium can result in low energy, lack of strength and muscle cramping. The optimum daily intake of potassium for adults is not known, but a diet with sufficient fruits and vegetables would provide about 3,500 mg per day.

- **Sodium** is needed to transport nutrients throughout the body and maintain normal fluid balance. It is also involved in healthy muscle functioning and it supports the blood and lymph systems. Too much sodium can raise blood pressure; however, the recommended daily amount of sodium from all sources for healthy adults is 3,000 mg or less.

Options for Drinking Water
in the Workplace

*R*emember those old-style water stations with those big blue plastic bottles? Did you know that those bottled water jugs and water-cooler reservoirs are susceptible to airborne contaminants and contamination from handling? When standard operations procedures are not followed, such as hand washing, disinfecting the bottleneck before changing the bottle or regularly sanitizing the water-dispenser reservoir, you increase the likelihood of contamination of the water cooler reservoir with harmful bacteria, viruses and algae.

Safer and cleaner options are bottleless reverse osmosis (RO) water coolers or dispensers, which, like RO for your kitchen tap, a water line is installed that runs through an RO system. Now you and your co-workers can enjoy clean uncontaminated drinking water. In addition, there are no bottle deliveries, reducing the carbon footprint; no bottles to change; no spills; no dry days; and no water cooler to sanitize and maintain. A bottleless RO water cooler or dispenser system can produce an unlimited supply of purified, great-tasting drinking water at a low fixed monthly price in any facility.

You can also bring your own filtered water to work. But what about those filters through which you pour regular tap water into a container? Carbon filters, such as Brita filters, are effective at absorbing chlorine and a part of other impurities; however, their efficiency begins to reduce after a month or so. They are also not effective for removing unwanted totally dissolved solids (TDSs).

What about bringing commercial bottled water in plastic containers to work with you? According to the International Bottled Water Association, $15 billion is spent annually on bottled water, with an annual average of 167 bottles being used per person.[14] However, the most alarming facts I found was that more than 80% of plastic bottles in the United States end up in landfill, and it takes up to 700 years for one plastic bottle to decompose. In the United States alone, some sixty million plastic bottles are used every day![15] So rather than adding to the problem, consider using a glass or stainless-steel bottle that can be refilled with your water or other cold beverage.

Now that you've raised the vibration of your drinking water by removing unwanted or harmful substances—or enhancing your water with minerals and balancing the pH, let's take a look at some other occasions when you use water.

Eco-Friendly Washing for Clothes and Dishes

What Does Eco-Friendly Mean?

*T*o me, eco-friendly means living in a manner that is beneficial to you and that does not harm the environment; in other words, living in harmony with our planet. There are many ways to be eco-friendly, from recycling, reusing and reducing consumption to conserving natural resources to being more energy efficient to reducing pollution to reducing our carbon footprints and more.

Every day we spend money on products, but do you ever stop to think about how these products affect the environment? Where does the packaging end up? Was the product imported? Was it made or grown locally? Is it made of synthetic materials? Does it contain preservatives, chemicals or GMOs? Can I compost it? These are questions you need to ask yourself before every purchase.

Besides grocery shopping, other high-ticket items in our household budgets are cleaning products. Let's compare the advantages and disadvantages of commercial cleaning products with eco-friendly and homemade options as cleaning products can significantly impact your living space, air quality, respiration, skin, internal organs and even your cells.

Washing Clothes

Laundry is a weekly event for most people. But did you know that most commercial laundry detergents are loaded with potentially toxic chemicals that could harm you, your family and the environment? Have you read the product label of your favorite laundry detergent lately? Most ingredients listed are either difficult to pronounce or vague, such as the ones listed below in a popular brand:

- Cleaning agent
- Buffering agent
- Stabilizer
- Brightening agent
- Fragrance

Those are pretty vague, aren't they? Now look at the following list of some of the ingredients commonly used in laundry detergents:

- **Linear alkyl sodium sulfonates (LAS):** This is a very strong synthetic cleansing agent, called an anionic surfactant on labels. To produce this substance, carcinogenic and reproductive toxins such as benzene are released into the environment. They also biodegrade slowly, making them even more hazardous to the environment.
- **Petroleum distillates (aka napthas):** These are chemicals that have been linked to cancer and can cause inflammation and damage the lungs and mucous membranes.
- **Nonyl phenol ethoxylate:** This is a common surfactant used in laundry detergents. It was found to biodegrade into even more toxic compounds that can stimulate breast cancer growth and feminize male fish. Phenols can cause death

in those who are hypersensitive to it and cause serious side effects at very low exposures. It is rapidly absorbed and can affect the whole body, in particular the central nervous system, heart, circulatory system, lungs and kidneys.

- **Optical brighteners:** These chemicals make clothing appear whiter, but they don't actually play a part in making them cleaner. These synthetic chemicals convert UV wavelengths into visible light, and some people experience allergic reactions when there is residue on their skin and they are exposed to sunlight. They are also toxic to fish and cause bacteria to mutate.

- **Phosphates:** These chemicals help soften water by removing minerals, and they prevent dirt from attaching to the clothes again while being washed, making the laundry detergents more effective. However, when phosphates get into the environment, they create imbalances in ecosystems by stimulating the growth of certain marine plants. Because of this, nowadays you can find "low-phosphate" or "phosphate-free" laundry detergents.

- **Sodium hypochlorite (household bleach):** A precursor to chlorine, this chemical has been responsible for more poisonings in the home than any other. In the environment, it reacts with organic matter, creating carcinogenic and toxic compounds that disrupt reproductive, endocrine and immune systems.

- **EDTA (ethylene-diamino-tetra-acetate):** This compound is used instead of phosphates to soften hard water and to prevent the activation of bleaching agents in the container before they're put in water. It does not degrade easily and can redissolve toxic heavy metals in the environment, allowing them to reenter the food chain.

- **Artificial fragrances:** Many of these are made from petroleum and do not degrade in the environment. They are toxic to fish and mammals, and they can cause allergies and irritation to skin and eyes.

Safe Laundry Detergents

Soapnuts
Soapnuts grow on trees of the genus *Sapindus* in warm and tropical regions of the world. Because the berry-like fruit contains saponins, which are a natural surfactant, they are used to make natural soaps. Detergents made from the soapnut berries are gentle, biodegradable, antimicrobial and safe to use in modern washing machines.

Homemade Laundry Detergent
Here is an inexpensive recipe for making your own laundry detergent:

1 cup (250 mL)	washing soda
1 cup (250 mL)	borax
1 bar	Dr. Bronner's pure castile soap, grated

Put all the ingredients in a food processor and pulse to combine well. Place in an airtight container and, voilà! For each load of laundry, use 1 tablespoon (15 mL) of the mix, less if you have a front-load machine.

Fabric Softeners and Dryer Sheets
Although they may make your clothes feel soft and smell fresh, fabric softeners and dryer sheets also contain toxic ingredients such

as chloroform, benzyl acetate and pentane, which are known to cause cancer and damage to the lungs, brain and nervous system. When clothes are heated in a clothes dryer or during ironing, the chemicals are even more dangerous. Toxic fumes fill our homes and go into the environment outside. Fabric-softener chemicals are designed to stay in clothes and slowly release over a long period of time. This slow release into the air affects the health of those wearing the clothes and of the people around them. Our sheets, towels and clothing are coated with these chemicals, and guess where they end up—they are absorbed through our skin, the largest organ of the body.

Here is a list of just some of the chemicals found in fabric softeners and dryer sheets:

- **Benzyl acetate:** Linked to pancreatic cancer
- **Benzyl alcohol:** Upper respiratory tract irritant
- **Ethanol:** On the Environmental Protection Agency's (EPA) hazardous-waste list and can cause central nervous system disorders
- **Limonene:** A known carcinogen
- **A-terpineol:** Can cause respiratory problems, including fatal edema, and central nervous system damage
- **Ethyl acetate:** A narcotic on the EPA's hazardous-waste list
- **Camphor:** Causes central nervous system disorders
- **Chloroform:** Neurotoxic, anesthetic and carcinogenic
- **Linalool:** A narcotic that causes central nervous system disorders
- **Pentane:** A chemical known to be harmful if inhaled

Safer Alternatives

Add ¼ cup (60 mL) of baking soda to the wash cycle to soften fabric.

Add ¼ cup (60 mL) of white vinegar to the rinse cycle to soften fabric and eliminate cling.

Check out your local health food store for a natural fabric softener that uses a natural base such as soy instead of chemicals.

Washing Dishes

Most commercial dishwasher detergents and liquid dishwashing products have a similar fate as laundry detergents. The chemicals they contain ultimately flow into our water supply and septic systems, affecting our environment. The question you need to ask yourself when buying household cleaners is, "Do I want to use what is convenient or do I want to use what is safe—for me, my family and the environment?"

Homemade dishwashing detergents work just as well as commercial ones, and they are more economical. Another benefit of homemade dishwashing detergent is the environmental impact— they don't contain harmful chemicals, and, furthermore, they eliminate plastic waste. (Remember, those plastic containers take hundreds of years to break down in landfill sites.)

Homemade Dishwasher Detergent Ingredients

To make eco-friendly dishwashing products you need a combination of the following ingredients:

- **Borax:** The technical name for Borax is *tetraborate*, a naturally occurring mineral.
- **Washing soda:** Washing soda is a natural, safe, chemical-free substance that is made from limestone and salt.

- **Kosher salt:** Kosher salt is 100% pure salt. In a homemade dishwasher detergent, it provides effective scouring qualities and eliminates tough stains from your dishes.
- **Citric acid:** Citric acid is made up of extracts from citrus fruits. It is a pure ingredient that adds fragrance to eliminate food odors; this is an optional ingredient.

Two Dishwasher Detergent Recipes

Recipe #1

½ cup (125 mL)	liquid castile soap
½ cup (125 mL)	water
1 teaspoon (5 mL)	freshly squeezed lemon juice
3 drops	tea tree extract
¼ cup (60 mL)	white vinegar

Put the castile soap and water in a bowl and stir to combine. Add the lemon juice, tea tree extract and vinegar and stir until well blended. Store in a squeeze bottle. Use 2 tablespoons (30 mL) per wash in a standard-sized dishwasher.

Note: Do not substitute conventional liquid soap for the castile soap unless it is a "low-sudsing" soap. Regular soaps create too many suds and cause the dishwasher to overflow.

Recipe #2

1 cup (250 mL)	borax
1 cup (250 mL)	washing soda
½ cup (125 mL)	kosher salt
½ cup (125 mL)	citric acid

Put all of the ingredients in a container and mix well. To use, just place 1 tablespoon (15 mL) of the powder into the dishwasher for each load.

Note that if you want sparkling clean dishes, just add white vinegar to the rinse dispenser; in addition to making your dishes sparkle, it takes away any leftover odors as well, and it helps to keep your dishwasher clean.

Two Hand Dishwashing Liquid Detergents

Recipe #1

8 cups (2 L)	water
1 (4-ounce / 115 g)	bar of soap, grated, or 4 ounces (115 g) plain soap flakes, natural and unscented
½ teaspoon (2.5 mL)	essential oil, any scent you want

Heat the water until it just begins to steam. Remove it from the stove and add the soap and essential oil. Stir to combine well, and then let sit for about 8 hours. Once the mixture has cooled thoroughly, pour the mixture into any reusable container.

Note: If using a bar of soap instead of soap flakes, freeze the bar beforehand to make it easier to grate.

Recipe #2: Old-Fashioned Hand Dishwashing Liquid Detergent

1 gallon (4 L)	water
2 cups (500 mL)	grated natural, unscented bar soap or soap flakes

Pour the water into a large pot and stir in the soap. Place on the stove and heat over medium-high heat until the mixture begins to boil. Keep stirring until all the soap has melted, and then lower the heat and simmer for about 10 minutes. Remove from the stove and let it cool, and then pour the liquid detergent into your chosen container. You need only about 1 teaspoon (5 mL) of the liquid for each sink full of hot water.

Living in Harmony

Whether directly or indirectly, the more chemicals we use to wash our clothes and dishes, the greater the chance that they find their way into our bodies, our homes and the environment. We can reduce this toxic load simply by using natural ingredients.

There is an intrinsic satisfaction I get from knowing I am part of an eco-friendly solution and not a contributor to a problem. Detoxing your home and your life and living in harmony can be a very simple task. Don't you agree?

High-Vibration Bath Formulas

*H*ave you ever come home from a hectic day at work or a stressful situation and just wanted to jump into the shower or slip into a warm bath to "wash your worries away"? Why do we gravitate to water to ease our worries? Perhaps it's because being in water makes us feel "lighter"; soaking in a warm bath is like returning to the womb—it's a safe and peaceful place.

Water is an element associated with cleaning and purification. Water also acts as a conduit to transport us from one state to another. On an energetic level, it is natural for us to gravitate to the water element to remove negative energy—either emanating from within us (thoughts, emotions), from other people's energy or from the environment around us. Water also helps cleanse and clear our auras, which are electromagnetic energy fields that surround our bodies and act as filters between our outer and inner worlds.

The health of your aura directly or indirectly affects the people around you—whether at home or at work. When your aura is weak, it is easy for you to take on the emotions, problems and stresses of others, which often leaves you feeling heavy, tired or agitated. When your aura is strong and evenly spaced around your body, you feel centered, solid and more whole; it is easier to get in touch with your emotions and be observant to your surroundings without taking on "other people's stuff."

Another energy system that can be cleansed with water is the chakra system. Chakras are spinning vortexes of energy that funnel universal energy into the body. Chakras are the organs of the body's energy system. Just as each of the body's organs has a purpose for maintaining good health, each chakra has a purpose for maintaining good health. When chakras are over- or underactive, they are unbalanced and do not feed the body the right energy to maintain optimum health. We can rebalance the chakras by removing negative energy. An effective way to do this is by taking a detoxifying or high-vibration bath using certain ingredients.

Bath Formulas

Below are a few high-vibration bath formulas I have learned over the decades that rebalance, remove negative energy and raise personal vibrational frequency. I encourage you to make time at least once a week for a high-vibration bath—your body and soul will thank you.

Note: for the following high-vibration baths, try to soak in the tub for at least 10 minutes, preferably for 20 minutes. Also immerse your entire body, including your hair for maximum effect.

Formula #1

The vinegar in this formula helps improve blood circulation in the skin, and, at the same time, works to alter the pH level of the skin.

1 cup (250 mL)	apple cider vinegar
1 teaspoon (5 mL)	salt

Combine the vinegar and salt and mix well. Add the mixture to a half tub of water and soak for 20 minutes. With conscious intention, release your worries and any negative influences around you. When you unplug the tub, visualize those worries draining away.

Formula #2

The baking soda added to your bathwater in this formula makes the water alkaline, which helps reinstate the acid-alkaline (pH) balance. This bath is beneficial not only for cleaning but also, over a period of regular use, for ridding the body of odor, lightening the aura and increasing a sense of well-being.

½ cup (125 mL)	baking soda
Regular amount	water in the bath

Add the baking soda to your bath water and swish until it dissolves. Enjoy!

Formula #3

This bath formula helps release physical and psychic tension, and it promotes a feeling of well-being. The combination of Epsom salts and sea or mountain salt facilitates the elimination of toxic substances from the body by means of stimulating the movement of fluids through the body's tissues.

¼ cup (60 mL)	Epsom salts
1 cup (250 mL)	baking soda
1 teaspoon (5 mL)	salt (sea salt or Himalayan mountain salt)

Dissolve all the ingredients in the bathwater and soak for 10 to 20 minutes.

Formula #4

The oat straw in this formula helps augment the skin's metabolism, which, in turn, facilitates detoxification of the body.

2 quarts (2 L)	water
1 handful	oat straw

Pour the water into a pot and stir in the oat straw. Place over medium-high heat and let boil for 25 minutes. Strain the liquid and pour it into a bathtub filled with hot water.

Herbal Bath Formulas

Herb baths are made by making tea with the herb. Simply place 1 teaspoon (5 mL) of the herb in a cup of boiling water. Let it steep and cool to room temperature, then strain out the herb and pour the cup into a half tub of water. Herbs possess therapeutic properties that can soothe your skin, relax muscles and joints and stimulate the circulation.

- **Herbal bath with basil:** Basil has cleansing and protective properties; it removes negative energy and reduces any further influences or accumulation of negatively. It's also effective if you deal with overly aggressive people or feel threatened or bullied by others. Basil also contains antibacterial properties and can be used to relieve skin problems and irritations.
- **Herbal bath with cinnamon:** Use a cinnamon stick, not ground cinnamon, for making this tea. Cinnamon helps bring calmness to you, and it calms those you interact with, helping to reduce quarrels and dissonance. Cinnamon is also believed to have a favorable influence on personal finance or income.

- **Herbal bath with nutmeg:** You can use ground nutmeg for making this tea. Nutmeg helps give one strength and confidence, especially when dealing with apprehensive people, during periods of personal stress or when called upon to speak in public.

- **Herbal bath with ginger:** Take a piece of ginger no bigger than your thumb and chop it into small pieces. Add to water and boil, then remove from the heat and set aside to steep. Strain and pour into the tub. Ginger has many curative properties that result in perspiration and, at the same time, augment circulation. Ginger also has the ability to invigorate as well as pull out toxic substances through the surface of the skin.

Your Weekly Homework

Get a "do not disturb" sign, affix it to your bathroom door and slip into a high-vibration bath!

House Cleaning au Naturel

One of my mother's most popular mantras was "convenience kills." She was a firm believer that it was healthier to cook from scratch, make your own baby food from fruit and vegetables and clean your home using natural products. She also taught me how to live on a shoestring budget and to "waste not, want not." (She was a very smart lady indeed!)

So, naturally, when I became a homeowner, I remembered those values from childhood, and I found simple and natural ways keep my home free of commercial chemical-based products. It's very convenient to grab a cleaning product off the grocery shelf, but it's not easy for most of us to read the label on the products we are buying. What are we putting in the air and on our floors, counter tops and furniture? What are the long-term health implications of using synthetic chemicals in the home or office? What about the environmental pollution caused by manufacturing and disposal of these commercial cleaning products?

A Natural Solution

You can detox your house by putting non-toxic cleaning products on your shopping list. Better yet, why not save money by making your own all-purpose cleaners using natural and safer ingredients? On the following page is a list of common environmentally safe products that can be used alone or in combination for a wealth of household applications:

- **Baking soda:** For softening water, cleaning, deodorizing and scouring.
- **Unscented soap:** Avoid soaps that contain petroleum distillates. Liquid, flake, powder and bar forms of soap are biodegradable and clean just about anything.
- **Lemon:** A strong acid that is effective against most household bacteria.
- **Borax (sodium borate):** Borax is a natural substance, is not carcinogenic and does not accumulate in the body or absorb through the skin. It can be used to soften water, clean, deodorize and disinfect, and it can be used on wallpaper, painted walls and floors.
- **White vinegar:** Use to cut grease and remove mildew, odors, wax buildup and some stains.
- **Washing soda (sodium carbonate decahydrate):** This is a mineral that can cut grease, remove stains, soften water and clean walls, tiles, sinks and tubs. It shouldn't be used on aluminum, and it can irritate mucous membranes.
- **Ethanol or 100 proof alcohol solution in water:** This is an excellent disinfectant that should be used instead of isopropyl alcohol, which was once recommended but is now thought to contribute to illness.
- **Cornstarch:** Use cornstarch to clean windows, polish furniture and shampoo carpets and rugs.
- **Citrus solvent:** Use to clean paint brushes and remove oil, grease and some stains. Note: This solvent may irritate skin, lungs or eyes in people with multiple chemical sensitivities.

Home Cleaning Formulas

All-Purpose Cleaner

½ cup (125 mL)	white vinegar
¼ cup (60 mL)	baking soda, or 2 teaspoons (10 mL) borax
2 quarts (2 L)	water

Stir the vinegar and baking soda or borax into the water and store in a container of your choice. Use to remove scale and stains on shower panels, chrome fixtures, windows and mirrors.

Air Fresheners

- Baking soda or vinegar mixed with lemon freshly squeezed juice can be put in small dishes that you place around the house to absorb odors.
- House plants reduce odors
- Put 1 tablespoon (15 mL) of vinegar in 1 cup (250 mL) water and simmer on the stove while cooking to prevent cooking odors.
- Use vinegar to wipe smells, such as from fish and onion, from utensils and cutting boards, and then wash in soapy water.
- Place bowls of fragrant dried herbs and flowers in a room.

Carpet Cleaners

- To remove carpet stains, mix equal parts of white vinegar and water in a spray bottle. Spray directly on the stain and let sit for several minutes. Clean with a brush or sponge using soapy water.

- To remove grease stains, sprinkle cornstarch on the spots and wait for 15 to 30 minutes before vacuuming.
- For a heavy-duty carpet cleaner, combine ¼ cup (60 mL) each of salt, borax and vinegar and mix to form a paste. Rub the paste into the carpet stains and leave for a few hours, and then vacuum.

Drain Cleaners

Mix ½ cup (125 mL) of salt in 1 gallon (4 L) of water, and heat, but do not bring to a boil, and then pour down the drain. This will provide a light cleaning.

For stronger cleaning of a clogged drain, pour ½ cup (125 mL) of baking soda down the drain, and then pour in ½ cup (125 mL) of vinegar. The resulting chemical reaction breaks down the fatty acids in soap and glycerine. After 15 minutes, pour boiling water down the drain. Caution: Use this method only with metal plumbing, not with plastic pipes. Nor should you use this method after using a commercial drain opener as a reaction with the vinegar can create dangerous fumes.

Floor Cleaners and Polish

Most floors can be cleaned using a vinegar-water solution. Following are some formulas for a variety of surfaces:

- **Vinyl and linoleum:** Mix 1 cup (250 mL) of white vinegar and a few drops of baby oil in 1 gallon (4 L) of warm water. For tough jobs, stir in ¼ cup (60 mL) of borax. (Use sparingly on linoleum.)

- **Wood:** Combine vegetable oil and white vinegar in a 1:1 ratio and apply a thin coat to the floor. Damp-mop a wood floor with equal amounts of white vinegar and water mixed with a few drops of pure peppermint oil.
- **Painted wood:** Mix 1 teaspoon (5 mL) of washing soda in 1 gallon (4 L) of hot water.
- **Brick and stone tiles:** Mix 1 cup (250 mL) white vinegar in 1 gallon (4 L) of water. Rinse with clear water.

Furniture Polish
For varnished wood, combine a few drops of lemon oil with ½ cup (125 mL) of warm water in a spray bottle. Shake and spray onto a slightly damp cloth and wipe the furniture. Wipe once more with a dry soft cotton cloth.

Removing Lime Deposits
Deposits in your kettle can be removed by combining ½ cup (125 mL) of white vinegar with 2 cups (500 mL) of water in the kettle and gently boiling for a few minutes. Rinse well with warm water while the kettle is still warm.

To remove lime scale from bathroom fixtures, squeeze lemon juice on the deposits and let sit for several minutes, and then wipe clean with a wet cloth.

Removing Mold and Mildew
Apply white vinegar or lemon juice full strength with a sponge or scrubby.

Mothball Substitute
Mothballs are made of para-dichlorobenzene (p-DCB), which harms the liver and kidneys. Aromatic cedar chips, sometimes

called *juniper*, wrapped in cheesecloth or cedar oil applied to an absorbent cloth will repel moths naturally. Moth-repelling sachets can also be made with lavender, rosemary, vetiver and rose petals. Another deterrent is dried lemon peel, which you can scatter in a clothes chest or tie in cheesecloth to hang in a closet.

Rust Remover
Sprinkle a little salt on rust and squeeze a lime over it to soak the salt. Leave for 2 to 3 hours and then use the leftover rind to scrub the residue away.

Scouring Powder
To clean surfaces such as the stove top and refrigerator, which shouldn't be scratched, apply baking soda with a damp sponge.

Shoe Polish
With a thick cotton or terry rag, apply a mixture of olive oil with a few drops of lemon juice to shoes. Leave for a few minutes, and then wipe and buff with a clean, dry rag.

Toilet-Bowl Cleaner
Combine ¼ cup (60 mL) with 1 cup (250 mL) of white vinegar and pour into the toilet bowl. Leave for a few minutes, and then scrub with a brush and flush to rinse. Alternatively, a mixture of 2 parts borax and 1 part lemon juice can be used.

Tub and Tile Cleaner
For basic cleaning, rub in baking soda with a damp sponge and rinse with fresh water. For tougher jobs, wipe with vinegar first, and then scour with baking soda.
    Note: Use vinegar sparingly as it can break down tile grout.

Wallpaper Remover
Combine equal parts of white vinegar and water and apply with a sponge to the wallpaper. This will soften the adhesive. Peel off the paper, applying more of the mixture to stubborn patches.

Window Cleaner
Mix 2 teaspoons (10 mL) of white vinegar and 4 cups (1 L) of warm water in a spray bottle. Clean windows with crumpled black-and-white newspaper (not colored) or a cotton cloth, and do not clean windows when the sun is shining on them or if they are warm as this will cause streaking. It is important not to use more vinegar than indicated, or it can damage the glass and eventually cloud it.

A Cleaner Greener Workspace
Fortunately, there are a growing number of green and eco-friendly commercial cleaning products available for companies to choose from. I encourage you to share this information at your workplace—starting with the human resources department as they have a vested interest in ensuring employees are healthy and productive.

You will find more information in the resources section at the back of the book.

Closing Words:
Water and Frequency

Water is the most critical resource issue of our lifetime and our children's lifetime. The health of our waters is the principal measure of how we live on the land.
~ Luna Leopold

*D*id you know that in virtually every glass of water we drink, some of the water has already passed through fish, bacteria, rocks, trees, worms in the soil and many other organisms, including people? Energetically, water has memory—of its source, its travels and its uses, and, therefore, the more recycled and unfiltered the water is, the lower its frequency as its purity diminishes.

The good news is, we can increase the frequency of our water supplies by filtering it, rebalancing its pH, purifying it and remineralizing it. Once we optimize the frequency of our water for consumption, we can enjoy the full benefits of hydration, detoxification and cellular regeneration.

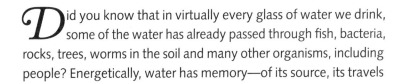

We can raise the vibration of the water we use for cleaning our homes and work spaces by reducing the use of chemicals, dyes, synthetic fragrances and other additives and replacing them with non-toxic, natural ingredients that will not harm us or our environment.

Lastly, we can increase the vibration around us by keeping our auras and energy systems clear, clean and "light" with a practice of taking regular detoxifying and spiritual baths, which allow us to remain centered and act with more clarity, which, ultimately, has a positive effect on our interactions with others.

Part IV

∞

Earth's Gifts

Himalayan Salt Crystal Lamps

*H*imalayan salt crystal lamps originated many millions of years ago in the Himalayan foothills of Pakistan. These salt crystals are said to be the purest in the world. The large crystal rocks are carved to make salt crystal lamps, in which an electric light bulb or a candle can be placed. These soft glowing lamps are unique as they come in various colors, shapes and sizes and provide an enchanting ambience to any room. They also refresh the air by emitting negative ions. I have used them for years, placing them in the bedrooms and my home office, where they naturally harmonize the air.

Positive Ions and Electric Smog

You can hardly find a place in the twenty-first century where there isn't wireless technology, electric devices and other modern-day tools designed to keep us connected to work, friends and even entertainment. But do you realize these technologies create a phenomenon known as *electric smog*, and that they are also a source of harmful positive ions? This electric smog worsens the quality of the air we breathe, which then adversely affects our physical and emotional states. Sources of these harmful positive ions, or electric smog, include computer monitors, vacuum cleaners, TV sets, microwave ovens, electric heaters, clothes dryers, dishwashers and tobacco smoke.

One alternative is to generate negative ions to counteract the positive ones, thereby harmonizing your air.

Negative Ions and Himalayan Salt Crystal Lamps

Himalayan salt crystal lamps are known to be natural negative-ion generators. The heat from a lighted salt crystal lamp attracts moisture, which evaporates through the salt, emitting negative ions in the process. How many negative ions a salt crystal lamp or candle holder can release depends on its size and how much heat the bulb or candle generates.

The negative ions work by creating biochemical reactions in your bloodstream that cause an increase in your serotonin levels—the brain chemical that makes you feel good. In addition to helping you to feel good, here are some other benefits of emitting negative ions into your living spaces:

- Increases chi (life energy)
- Eliminates allergens
- Aids in the relief of hay fever, sinus-related symptoms and asthma
- Reduces the effects of seasonal affective disorder (SAD)
- Supports the immune system
- Aids in clearing bronchial passages
- Improves indoor air purity and odors
- Improves mental alertness
- Promotes general well-being

Himalayan salt crystal lamps can be placed in spas and bathrooms; however, too much humidity can dissolve the crystals quickly and reduce their effectiveness for producing negative ions. Do turn your salt crystal lamps on every day to prevent any moisture buildup as well as to keep your rooms filled with negative ions.

You can use large salt crystal lamps for use at home in large rooms to ensure sufficient negative ions are emitted in the room, but for office use, you can obtain a lamp the size of a night-light or small globes with a USB key, which is beneficial for placing near your computer.

Taking Care of Your Himalayan Salt Crystal Lamp

Because salt crystal lamps attract moisture, it is best to store them in a dry environment. If you place your lamp on a windowsill or table, make sure you place a waterproof saucer or other base under the lamp so that any excess moisture it attracts does not absorb into your furniture. Over time, as the lamp constantly attracts moisture and then dries out, it tends to acquire a crystalline buildup on the surface. When this occurs, simply wipe the lamp with a damp cloth.

How Crystals and Gemstones Can Influence Your Home and Work Spaces

*M*other Nature has supplied us with a bounty of gifts in the form of crystals and gemstones. They offer numerous benefits, physically, metaphysically and environmentally. In this section, we will explore various benefits of using crystals and gemstones at home and at work and learn about the positive ways you can transform your life by creating a crystal grid. You can purchase these gifts from the earth at metaphysical stores or rock and gemstone outlets.

Crystals and gemstones can be worn or placed near you in a room on a desk or workstation, even on your car's dashboard. It is important remember that crystals and gemstones are living tools, and from time to time they need to be cleansed as they absorb, transmute and amplify certain energies. I try to clean my gemstones every four to six weeks, depending on the environment they are placed in. A few ways you can cleanse your crystals and gemstones include a technique called *smudging* (burning sage or sweetgrass), washing them in water infused with some Himalayan salt (provided they are not porous), placing them in the sun for a few hours or putting them directly in the earth overnight. There are hundreds of crystals and gemstones you can place in your home or workspace.

Below are just a few you can consider, along with their respective metaphysical properties and influences at home or at work. Have fun with them, and learn about other crystals in the suggested reading list at the back of the book.

Benefits of Using Crystals and Gemstones at Home and at Work

Name of Crystal or Gemstone	Influences and Benefits at Home	Influences and Benefits at Work
Blue Lace Agate	A peacemaker; minimizes family confrontations	Helps you think on your feet, especially when called to give speeches or presentations
Amazonite	Helps to reduce anger and irritability; also good for preventing you from bringing your work problems into the home	Helps you leave work problems at work; aids in success at work by helping you to stay focused
Amber	A protective stone; helps guard against electric smog/techno-pollution	Reduces confrontational attitudes in others
Amethyst	Helps soothe anger and impatience	Helps soothe anger and impatience
Carnelian	Attracts abundance to your home and family	Place on your desk to radiate positive energy
Celestite (blue)	Helps break patterns of negativity	Improves communication, namely the ability to hear what is being said

Name of Crystal or Gemstone	Influences and Benefits at Home	Influences and Benefits at Work
Chrysocolla	Helps at home to heal after a traumatic incident	Improves communication with difficult people
Citrine	Helps clear negative energy	Helps clear negative energy
Emerald	Attracts abundance, money and prosperity	Helps prevent you from being overburdened with responsibility
Fluorite (green)	Helps bring order to chaos; stabilizes family dynamics	A stone of cooperation; brings order to chaos and support when goals need to be accomplished
Garnet (red)	A stone of protection as well as a psychic shield	Helps you get recognition for the work you have done
Hematite	Stone of self-healing; helps you overcome irrational fears, such as of flying or heights	Keep this stone close when you are involved in legal issues or disputes
Jade	Attracts good fortune into the home	Helps attract new business; helps reduce interaction with critical people
Kyanite	Brings calmness to the home, even in the midst of adverse or frustrating events	Helps you remain cool and calm; a stress-buster

Name of Crystal or Gemstone	Influences and Benefits at Home	Influences and Benefits at Work
Lapis Lazuli	Fosters loyalty, family unity and contentment	Fosters trust and helps you maintain your integrity
Malachite (green)	A cleansing crystal; helps reduce noise and computer and light pollution	A cleansing crystal; helps reduce noise and computer and light pollution
Obsidian (green)	Helps overcome the feeling of stagnation	Helps overcome the feeling of being stuck in a rut
Peridot	Attracts good fortune and abundance	Attracts new business and growth
Quartz (clear)	Promotes harmony in relationships; brings a sense of optimism	Transmutes pessimism into optimism
Scolecite	Improves tolerance and lessens violence within the home or the family	Promotes tolerance in the workplace and improved understanding
Tiger Eye (brown/gold)	A stone of luck and prosperity	Helps attract financial success
Topaz (blue)	Keeps communication lines open, even when things get hectic	Keeps communication lines open despite chaos or crisis
Tourmaline (black)	Helps you focus when there are distractions	Brings focus and order to chaos
Zinc	Helps you cope with sudden or unexpected change	Helps you cope with and solve sudden problems

Crystal Grids

Introduction to Crystal Grids

Crystals have the unique ability to transmute energy from negative to positive. They also have a natural capacity to amplify energy indoors and outdoors. A crystal grid is formed when crystals are placed in a particular geometric pattern for the specific purpose of directing energy toward a goal. The stones or crystals are then charged by your intention and energy.

A grid can be used for a variety of goals, from helping to solve a problem to breaking a negative habit to attracting abundance or prosperity to fostering peaceful, joyful and loving relationships to facilitating healing on many levels to acting as a shield or protection. Grids can also be used to charge food, water or clothing and change the energy field in a room or an entire building. Whatever your intention, the crystal grid is an effective use of crystal energy when done with a clear, pure intention. Grids can also be placed outside your home in your garden or along the edge of your property line.

In a crystal grid, all crystals are energetically connected and communicate with each other. The grid acts as a bridge between subtle intention and manifestation, where the crystals act as conduits. One thing to note about crystals is that they have a memory-holding capacity and will follow instructions for approximately seventy-two hours with accuracy. Although they attract light and energy, they are not able to discriminate between negative and positive light and vibration. For this reason, it is imperative that you cleanse and positively "charge" your crystals before placing them in the grid, which I describe on page 134 "Guidelines for Making a Crystal Grid."

Types of Crystals to Use and Their Properties

The crystals that really resonate with me for creating a crystal grid come from the quartz family. Quartz crystals are believed to be the most powerful amplifiers of energy. They enhance energy by absorbing, storing, balancing, focusing, transmuting and transmitting. Quartz also enhances thoughts, which are a form of energy. Because it directs and amplifies energy, it is extremely beneficial for manifesting, healing, meditation, protection and channeling. Due to its ability to balance energies, quartz is excellent for harmonizing and balancing one's environment, as well as relationships. Quartz is also good for energizing other crystals. The quartz family of crystals includes clear quartz, smokey quartz, rutilated quartz, tourmalinated quartz, rose quartz, citrine and amethyst.

- **Clear quartz:** A good healing stone, powerful in transmitting and transmuting energy.
- **Smokey quartz:** Beneficial for transmuting negative energies; a good grounding stone; almost acts like a vacuum to absorb and transmute energies and, therefore, requires regular cleaning.
- **Rutilated quartz:** Contains rutile inclusions within the stone and is nicknamed "angel-hair quartz" for this reason. These inclusions help magnify the energy of the quartz and make it a powerful conductor.

- **Tourmalinated quartz:** This quartz contains inclusions of black tourmaline. One property of tourmaline is its ability to protect against electromagnetic pollution, or electric smog. One drawback to this crystal is that it is often opaque or cloudy. Crystal grids work more effectively when the crystals are clear. The golden rule here is the clearer the crystal, the higher the vibration.
- **Rose quartz:** Brings a soft, peaceful energy of self-love and works well on the heart chakra. It is, however, the weakest member of the quartz family for amplifying energy or giving protection. Rose quartz also requires frequent cleaning.
- **Citrine:** Contains a fiery energy, and, like the rays of the sun that banish all darkness, it helps us overcome dark emotional states and keeps us positive. Citrine is also one of only two gemstones that is self-cleaning (the other is kyanite), which also makes it a great crystal for keeping vibrations high.
- **Amethyst:** A protective stone. It also aids in dreaming and spiritual practices. In ancient times, amethyst was believed to ward off evil and bad luck and help people overcome addictions. A meditation or healing room would be a wonderful space for an amethyst grid.

Some Basic Grid Formations and Their Applications

There are many ways to create crystal grids—probably more ways than there are days in a year. Some grids are simple circles, whereas others are based on traditions of sacred geometry or symbolic forms such as the Tree of Life, the Merkabah, the Ankh or the Infinity symbol.

The following is an introduction to five basic grid types. I encourage you to look at the suggested reading list at the end of this book for further study and exploration into the magical world of crystal grids.

Four-Sided, or Pyramid, Grid

This grid is a two-dimensional representation of a pyramid and uses five main stones, one in each of the four corners and one in the center. The four-sided crystal grid pattern creates a field of energy that projects a lovely field over a wide area. Anything placed within this field absorbs and stores or uses this energy, so you can make a large grid or put the grid on a box in which you keep things you want kept clear and charged. This is a very stable grid and enhances programs or intentions that concern long-term plans, healing, relationships, Earth healing and environmental work, among other uses.

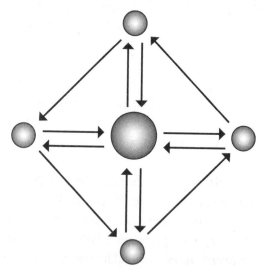

Six-Sided, or Star-of-Solomon, Grid

This is a grid with a center stone and six outer crystals or stones in a circle around it, and it can be used with or without linking the crystals. The links from or to the center create a Star of Solomon. It is a popular pattern often used for Reiki grids, distance healing and in world-peace healing grids. In metaphysics, the number six represents harmony and perfection.

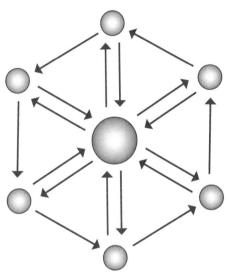

Relationship Grid

This is a grid pattern made of two groups of three stones set in triangles with the points toward the center. A center or "master" stone may be placed between the two triangles to act as an energy transmitter or to represent a particular issue or aspect of the situation being worked on. Alternatively, one stone may be used as the center for both triangles. The relationship grid, as its name reveals, is typically used to help resolve relationship issues.

A popular crystal that represents the heart chakra, namely rose quartz, is commonly used for this type of grid.

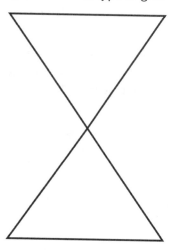

Crystal Love Grid

This grid, combined with your intentions, can assist you in healing past hurts, forgiving those who have trespassed against you and allowing you to receive unconditional love. It can also help you recharge a long-term relationship that may need an energetic "tune-up."

This love grid is designed to generate unconditional love all around you, whether it is for romantic love, self-love, a loving family environment or loving friendship. The suggested stones for this grid are five rose quartz crystals, five clear crystal points and one rose quartz ball as your "master stone," which you place in the center.

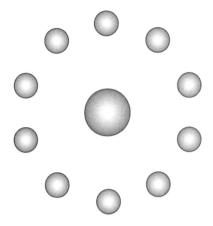

Prosperity Crystal Grid

This grid, combined with your intentions, can assist you in attracting abundance and promoting contentment. It uses the Infinity symbol to create endless abundance for success in your career and financial abundance, and for any part of your life that you are ready to transmute from a place of lack to a place of prosperity, where you can thrive. The suggested stones for this grid are six black tourmaline, six light or dark-green aventurine or emeralds and a citrine or carnelian cluster at the center.

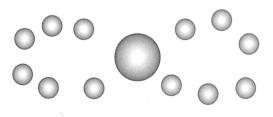

Guidelines for Making a Crystal Grid

1. Prepare your space and stones

I suggest you choose a space that you have easy access to—a place that you know the crystals will not be disturbed or displaced once it is established. It can be a table, an altar or even a shelf. Clean and clear the space either with incense, by smudging or with an instrument with a piercing sound such as a Tingsha, which is also effective in clearing stagnant, dark or heavy energy from a space. Also choose one of the elements to clean your gemstones, such as air (smudging), light (sunlight), water or earth as described at the beginning of this section.

2. Set your intention and purpose for the grid

The power of your intention is how you empower the stones' energies and charge the crystals. It is important to be clear on your intention.

3. Place the crystals and stones in a geometric pattern

See the section that begins on page 129, "Some Basic Grid Formations and Their Applications," which explains a few basic grid formations and why certain patterns are beneficial for different goals.

4. Charge and program the master crystal

The master crystal is the crystal that is placed at the center of the grid. A clear quartz is a popular choice because its energy frequency is neutral, and it can therefore be used for virtually any purpose you wish to program it for.

To program the master crystal, you consciously project thought forms into the crystal. It can be seen as being similar to computer programming, whereby information and commands are stored, but rather than in the hard drive of a computer, you are storing these in the molecular structure of the quartz crystal in the form of magnetically charged instructions. You can think of your projected thoughts as being pictures or other sensory information that you power or intensify with your emotions. Following are two types of programming you can use:

i. *Third-Eye Programming:* Sit quietly in an alert state and center yourself. With a calm mind and a positive intention, visualize your goal as having already been achieved. Now place the crystal between your eyebrows in the area of your third eye. Close your eyes and go inside your crystal. Picture the end result; see the manifestation of your goals unfold, making the picture as clear as possible, using as many of your senses as you can: See it, hear it, touch it, smell it, taste it. When it feels real, come out of your crystal and open your eyes.

ii. *Sense-Based Programming:* Sit quietly in a relaxed and alert state. Watch your breath flow slowly in and out, and allow your body to relax even more. Feel peace flow in as you inhale, and negativity leave you as you exhale. Hold the crystal in your dominant hand and place your other hand over it, and then ask that the crystal may be used only for the highest good of all concerned. Touch the crystal to your heart and allow loving feelings to flow from your heart. Feel this love coming from deep within as it pours into your crystal. Visualize an energy connection between you and your crystal, and then speak to the crystal aloud or silently. State your intentions, for example, "I intend for this crystal to be effective for _____ in light and love for the highest good of all concerned." Picture in your mind the end result, that which you want most, happening inside your crystal. Make the picture as clear as possible, using as many of your senses as you can: See it, smell it, touch it, hear it, taste it. Slowly open your eyes.

5. Charge the other stones in the grid
Now that the stones are in place and you have charged the master crystal, you are ready to energize the other stones in the grid. Place your master crystal in your dominant hand, and with the crystal point facing down above the grid, imagine energy pouring from the master crystal and charging your crystal grid. You can also charge the grid by drawing an imaginary line from one stone to the other using the master crystal.

6. Give thanks and repeat a closing affirmation or mantra
To complete the exercise, give thanks for all that you have received and for what is manifesting in the present moment, and then you can repeat an affirmation or mantra to conclude the process. One of my personal favorites, which I usually repeat seven times, is "I empower this grid with the highest vibration of light and love to heal."

Flowers, Happiness and Well-Being

*P*erhaps you like the color? Maybe it's the fragrance that captured your attention, or could it simply be the aesthetically pleasant visual break from looking at your computer screen, TV or cell phone? Whatever the attraction, flowers can enhance your home or work by adding natural charm and beauty.

Bringing flowers indoors also has a "happiness effect." Having flowers in the home or office soothes us, which naturally reduces stress, which, ultimately, causes us to feel more relaxed, secure and happy! Hundreds of studies have been conducted on the effect ornamental plants and flowers have on our mood, stress levels and behavior. One study conducted by the Norwegian University of Life Sciences and Uppsala University in Sweden found that potted plants in offices can reduce fatigue, headaches and stress, and that the more plants a worker could see from the desk, the fewer sick days they took.[1]

In a US study at Washington State University whereby workers were given timed computer tasks to complete in rooms with plants and then without plants, researchers found that the workers were 12% more productive and had quicker reaction times when there were plants in the room. They were also less stressed and their blood pressure was lower.[2]

Plants and flowers can also be healing. Have you heard of the term *horticulture therapy*? I like to think of it as nature taking care of us when we take care of nature. Some hospitals have introduced horticulture therapy to their patients. It's a technique whereby patients care for and nurture plant as a means to reduce their recovery time. Physicians observed that patients who physically interact with plants have a significantly reduced recovery time after medical procedures than those who don't interact with plants.[3,4]

Who knew that such small, unassuming flowers and ornamental plants can not only raise the vibration of your living space with fragrance, color and visual stimulation, but they also help raise your vibrational frequency? They do this by increasing the levels of positive energy, which helps us achieve a more positive outlook on life, counteract stress and facilitate healing.

What types of flowers do you like? Allow your intuition to guide you in selecting flowers that make you feel good—flowers that you are naturally attracted to. The greater your attraction, the more relaxing and soothing they will make you feel, and they will have a pleasant effect on your living and work spaces. If you are still uncertain about what to pick, try to look beyond their physical form, color and fragrance and consider selecting flowers based on their symbolic meaning.

Flowers and Their Meaning

The Language of Flowers

There is a language, little known,
Lovers claim it as their own.
Its symbols smile upon the land,
Wrought by nature's wondrous hand;
And in their silent beauty speak,
Of life and joy, to those who seek
For Love Divine and sunny hours
In the language of the flowers.
~ *From Wild Flowers of Canada* by
George Iles, 1875

The Symbolic Meanings of
Herbs and Flowers

Herb or Flower	Meaning
Aloe	Healing, protection, affection
Angelica	Inspiration
Basil	Good wishes
Bay	Glory
Black-eyed Susan	Justice
Carnation	Alas for my poor heart
Chamomile	Patience

Herb or Flower	Meaning
Chives	Usefulness
Chrysanthemum	Cheerfulness
Clover, white	Think of me
Daffodil	Regard
Daisy	Innocence, hope
Dill	Powerful against evil
Fern	Sincerity
Forget-me-not	Forget-me-not
Geranium, oak-leaved	True friendship
Goldenrod	Encouragement
Holly	Hope
Hollyhock	Ambition
Honeysuckle	Bonds of love
Horehound	Health
Hyacinth	Constancy of love, fertility
Hyssop	Sacrifice, cleanliness
Iris	A message
Ivy	Friendship, continuity
Jasmine, white	Sweet love
Lavender	Devotion, virtue
Lemon balm	Sympathy
Lilac	Joy of youth
Lily-of-the-valley	Sweetness
Marjoram	Joy and happiness
Mint	Virtue
Morning glory	Affection
Myrtle	The emblem of marriage, true love
Parsley	Festivity

Herb or Flower	Meaning
Poppy, red	Consolation
Rose, red	Love, desire
Rosemary	Remembrance
Sage	Wisdom, immortality
Salvia, blue	I think of you
Salvia, red	Forever mine
Sorrel	Affection
Sweet pea	Pleasures
Sweet William	Gallantry
Sweet woodruff	Humility
Thyme	Courage, strength
Tulip, red	Declaration of love
Violet	Loyalty, devotion, faithfulness

The Vibration of Flowers

Flowers not only carry certain meanings across different cultures, but they also add a positive vibration to a place. Perhaps that is why we bring flowers to people who are sick in the hospital and place them around altars of churches and in funeral homes. I love to observe the effects flowers have in a room, and over the years, I have used a variety of flowers for different social and business occasions. Following are the vibrational effects of a few of my favorite flowers:

- **White Roses** carry a vibration of love. These flowers work as a "vacuum," absorbing negativity from the surroundings. Roses are also beneficial to reduce or temper anger in a home.
- **Carnations** have a healing vibration, and their presence is very beneficial in reducing and removing emotional and

mental turmoil. They make wonderful gifts for those in hospitals.

- **Chrysanthemums** have a nurturing and nutritive vibration. They are a wonderful addition to a home when a new baby arrives. They are also very beneficial in a nursery.
- **Gardenias** have a vibration that promotes harmony between people, especially in business partnerships or marriages. These flowers absorb vibrations of discord and dissonance.

Gardening: Making a Therapeutic Connection

*A*nother beneficial way to appreciate and connect with nature is to spend time gardening, either outdoors or indoors. It doesn't matter whether you live in the country or the city, in an apartment or a house; there are many creative ways to enjoy gardening. Depending on your geographic location, climate and how much time you can spare to maintain a garden, here are just a few types of gardens you can create:

- Vegetable
- Flower
- Organic
- Container
- Water
- Indoor
- Herb
- Rock
- Japanese
- Raised bed
- Upside down
- Bird
- Roof
- Wall
- and my personal favorite—a community garden!

A few years ago we planted a herb and vegetable garden at our home. I carefully researched and found an heirloom seed company and purchased the herb and vegetable seeds that would grow best in our region. Why heirloom seeds? Heirloom seeds are defined as seeds that have been cultivated for fifty years or more, and they include seeds for vegetables, fruit and flowers. All heirloom plants are cultivated varieties (not wild) that have been deliberately selected for specific characteristics, such as flavor, texture, color, hardiness and yield. When grown and harvested correctly, those characteristics are retained from one generation to the next. In theory, the heirloom tomato that was grown by your great, great, great grandmother has the same leaf type as the one you grow in your garden today!

At the other end of the spectrum, genetically modified organisms (GMOs) are organisms, including seeds, that have been altered by engineering that changes their genetic makeup. This manipulation changes the organisms' characteristics, such as resistance to insects, herbicides and extremes in weather. Some GMO crops have been modified to manufacture their own herbicides, GMO corn is an example of this. Genetic manipulation can include the introduction of genes from other species; for example, DNA from fish, frogs and bacteria can be introduced to various plant species. There is controversy about the safety of consuming genetically modified food as well as the possibility of GM crops "polluting" neighboring nonGM crops in fields, and regulations vary from country to country.

The harmful effects of GMOs on our health include cancer, degenerative diseases, allergies, bacteria and superviruses, and their impact on our environment includes soil toxicity and soil and tree sterility. GMOs even impact insects and larger animals; an example of this is the slow but steady disappearance of bees and monarch butterflies. My personal solution to counteract this technology is to empower myself with knowledge about healthful alternatives and exercise my right—a basic human right—to safe food. I encourage you to explore the resources section at the back of this book to learn more about heirloom seeds and how you can obtain them and enjoy the same real foods that were grown many generations ago.

One of the greatest joys I get from gardening is simply the joy of nurturing and helping a living thing grow. Whether I am patiently watering the seeds in our mini-greenhouse, waiting for them to sprout, gently placing the little plants outside in the prepared garden bed or turning on the hose to quench their thirst on a hot summer day, I feel blessed to be part of their growth and grateful for what Mother Earth has provided.

Did you know that gardening is also therapeutic? Studies have shown that people who spend time cultivating plants have less stress in their lives. Like flowers, gardening has a soothing effect on us and uplifts our spirits. Gardening is also a great outlet to release pent-up frustration or negative emotions and allows us to focus on things of beauty, increasing our ability to have more joyful and loving emotions. Gardening can also have a therapeutic effect on people who have undergone mental or physical trauma. Nurturing a plant can provide these people with the means to work through their problems and heal their wounds, whether superficial or deep within. Gardening improves their mental states, putting them in a better place for healing.[5,6]

A wonderful example of how gardening can heal and transform lives comes from the work done by a charity based in the UK called "Thrive." Established in 1998, their mission is to use their passion for the power of gardening to change the lives of people touched by disability. Thrive offers programs and educational opportunities that allow disabled people an opportunity to improve their physical health through exercise and learning how to strengthen their muscles to improve their mobility and mental health through a sense of purpose and achievement, learn new skills and connect to others, which reduces feelings of isolation while building a community.

Earlier I mentioned that community gardens are my favorite type of garden. To me, they exemplify how people can come together with common goals and work in harmony for the greater good. A community garden also provides those without access to land or open spaces a chance to connect with the environment and grow their own food as well. Most gardens of these are tended communally, which allows everyone to help out and share the bounty of the garden. There are an estimated 18,000 community gardens in the United States and Canada. Please see the resources section at the back of this book for sources for locations of community gardens.

Closing Words:
Earth's Gifts and Frequency

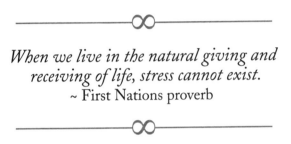

*When we live in the natural giving and
receiving of life, stress cannot exist.*
~ First Nations proverb

I've had the good fortune to attend many sunrise ceremonies
and sit with the elders and listen to their wisdom. I have heard
many stories, but the one that burns deepest in my heart is the one
about the promise each living thing (animals, plants and humans
alike) made to our planet. The promise humans made was to be
the Earth's keepers—to honor, cherish and protect the environment.
We have not kept our promise.

Everything we do to the environment, we do to ourselves.
When we pollute the oceans, lakes and rivers, we toxify our bodies;
when we pollute the land, we contaminate our food supply; when
we pollute the air, we have respiratory problems. We are vibrational
beings designed to be healthy, happy and live in harmony with
nature. The further we distance ourselves from the natural world,

including nature, natural foods, natural light and natural sounds, the more suffering and disease we experience. It's time to simplify our lives and reconnect with the land, the environment and nature. It's time to keep our promise and become Earth keepers. Are you ready?

You now possess the *why* and the *how* to raise the vibration around you. You have the power to choose the highest good for yourself *and* the environment, and you understand how interconnected and interdependent we are with nature. We have explored many aspects of working with the four elements to create healthy and harmonious living spaces, which, ultimately, improves our health and well-being, but there is one critical element we have not explored—the state of "being" in harmony with all aspects of life.

Part V

―――――∞―――――

The Ultimate Goal

"Being" in Harmony

*K*nowing what to do and *doing* what you know are two vastly different things. For instance what good is it to have a Himalayan salt lamp in your living room when you are still yelling at your children; changing the light bulbs in your home but not making time to sit and relax or walk outside to enjoy natural light; or growing vegetables in your garden but continuing to purchase processed food laced with aspartame, MSG and synthetic vitamins?

How do we shift from a state of *knowing* to a state of *being*? *It begins with having confidence in your awareness of what it takes to create harmony.* Have confidence in your decisions about what you choose to buy and use. Have confidence in your choice to purchase what is safe and healthful, not what is convenient or "popular." Have confidence in your decision about how you speak to others and interact with them at home and at work. Have confidence in how you choose to spend your time and money rather than live in the automatic pilot of consumerism and busyness.

Allow harmony to be the first goal in everything you do. Often when I am sitting in a café, I hear small groups of people talking about this and that with an incessant need to have the last word or make *the* most sensational comment, as if they are in a competition, not a conversation. Why not sit in peace and attend to what your friend is saying—with interest, with compassion? If someone has done something a little differently from you, why choose to compare and criticize their actions? Why not appreciate that they attempted to do something and completed it? When we allow our words and actions to be driven by our egos, we narrow our view of the world as we focus only on the "me," causing us to lose sight of the bigger picture, the bigger lesson or message.

Be present with those around you. This may sound like an insurmountable goal given all the technological distractions we have in society today—but it is achievable. Being present takes discipline. By discipline I don't mean something unpleasant or a chore that is imposed from outside. Rather, discipline is an organic process that expands naturally from within, driven by your preferences and your values. Being present means embracing a state of simplicity; put away the pacifiers (gizmos, gadgets, earplugs) and simply look, observe and appreciate the people around you, the people you live with and work with. At home, families have to make a collaborative effort to connect—maybe go on a techno-diet together (turn off your cell phones and stay away from Facebook for two hours when everyone is at home together). What would that be like? If you work in an office, it's easy and tempting to write an email and hit send, but I challenge you to get out of your comfort zone and walk over and *talk* to your co-worker.

Pay attention to the good deeds that others do and praise them as they do these deeds. Acknowledge the goodness in others—with a kind word, a compliment, praise, thanks. When you allow yourself to see goodness in others and in the world and feel goodness, a great sense of good health and wholesomeness fills your life, and you cannot help but project that goodness and healthiness to others.

Trust in yourself and your ability to choose the highest good at all times. Now you are equipped with the knowledge you need to create health and harmony in your home and workspace, but that knowledge has no power until you *apply it* to everyday living. It's time to make the vision of harmonious living a reality! I know you can!

I encourage you to continue seeking knowledge and start applying the principles of conscious living to every moment of your day. You can begin by reviewing the Suggested Reading and Resources sections at the back of the book.

References

Part I: Air

1. Dale Purves et al., eds., *Neuroscience, 2nd ed.* (Sunderland, MA: Sinauer Associates, 2001).
2. Ibid.
3. Ibid.
4. Dawn James, *Raise Your Vibration, Transform Your Life: A Practical Guide for Attaining Better Health, Vitality and Inner Peace* (Uxbridge, ON: Lotus Moon Press, 2010), 87.
5. Connie Higley and Alan Higley, *Reference Guide for Essential Oils*, 13th edition (Spanish Fork, UT: Abundant Health, 2012).
6. Bill C. Wolverton, Anne Johnson, and Keith Bounds, "Interior Landscape Plants for Indoor Air Pollution Abatement," NASA Report, (1989), http://ntrs.nasa.gov/archive/nasa/casi.ntrs.nasa.gov/19930073077_1993073077.pdf.
7. American Society for the Prevention of Cruelty to Animals, "Toxic and Non-Toxic Plants," (2012), www.aspca.org/pet-care/poison-control/plants.
8. SomaEnergetics, "What Are the Ancient Solfeggio Tones?" (2012), www.somaenergetics.com/solfeggio_frequencies.php.
9. John L. Carter and Harold Russell, "A Pilot Investigation of Auditory and Visual Entrainment of Brain Wave Activity in Learning Disabled Boys," *Journal of the Texas Center for Educational Research 4* (1993): 64–73.

10. Maya Ruvinshteyn and Leonard Parrino, "Benefits of Music in the Academic Classroom," (paper presented at the 16th annual conference of the New Jersey Faculty Development Network, April 2005).

11. Mei-Yueh Chang, Chung-Hey Chen, and Kuo-Feng Huang, "Effects of Music Therapy on Psychological Health of Women during Pregnancy," *Journal of Clinical Nursing 17* (2008), 2580–2587.

12. Sarah McKee, "Gentle Musings: Assessing Effects of Live Harp Music upon Patients, Family and Friends, and Staff at Massachusetts General Hospital's Cancer Center," *The Harp Therapy Journal* (Fall 2004).

13. Huffpost, "Harp Music Stabilizes Blood Pressure, Lessens Pain in ICU Patients: Study," Healthy Living (2012), www.huffingtonpost.com/2012/08/04/harp-music-blood-pressure-icu-patients-pain_n_1734615.html.

14. David Gutierrez, "Bird Songs Lend Therapeutic Powers to Hospital Patients," *Natural News* (December 4, 2010), www.naturalnews.com/030612_bird_songs_therapy.html.

Part II: Light

1. Wikipedia, "Phase-out of Incandescent Light Bulbs" (2013), http://en.wikipedia.org/wiki/Phase-out_of_incandescent_light_bulbs.

2. Igor Knez, "Effects of Colour of Light on Nonvisual Psychological Processes," *Journal of Environmental Psychology 21*, no. 2 (June 2001): 201–208.

3. John Nash Ott, "Influence of Fluorescent Lights on Hyperactivity and Learning Disabilities," *Journal of Learning Disabilities 9*, no. 4 (1976): 417–422.

4. John Nash Ott, Lewis W. Mayron, and Ellen L. Mayron, "Light, Radiation, and Academic Achievement: Second-Year Data," *Academic Therapy 11*, no. 4 (1976): 397–407.

5. Tatsiana Mironava, Michael Hadjiargyrou, Marcia Simon, and Miriam H. Rafailovich, "The Effects of UV Emission from Compact Fluorescent Light Exposure on Human Dermal Fibroblasts and Keratinocytes In Vitro," *Photochemistry and Photobiology 88*, no. 6 (2012): 1497–1506.

6. Victoria Ward, "Energy Saving Light Bulbs 'Contain Cancer Causing Chemicals,'" *The Telegraph* (April 20, 2011), www.telegraph.co.uk/health/8462626/Energy-saving-light-bulbs-contain-cancer-causing-chemicals.html.

7. U.S. Environmental Protection Agency, "Mercury Releases and Spills" (2012), www.epa.gov/hg/spills.

8. Val Jones, "Why Is McDonald's Yellow? The Role of Environment on Eating Behavior" (2008), http://getbetterhealth.com/why-is-mcdonalds-yellow-the-role-of-environment-on-eating-behavio/2008.11.04.

9. Lindsay Gruson, "Color Has a Powerful Effect on Behavior, Researchers Assert," *The New York Times* (October 19, 1982), www.nytimes.com/1982/10/19/science/color-has-a-powerful-effect-on-behavior-researchers-assert.html?pagewanted=all.

10. W. H. Eddy, L. Streltzoff, and J. Williams, "The Effect of Negative Ionization on Transplanted Tumors," *Cancer Research 11* (1957): 245.

11. Homer S. Black and Elizabeth W. Rauschkolb, "Effect of Light on Skin Lipid Metabolism," *Journal of Investigative Dermatology 56* (1971): 387–391.

12. Solar Healing Center, "Sun-Gazing History: Process" (2003), http://solarhealing.com/process.

13. David E. Blask, "Melatonin: Chronobiological and Chronopharmacological Role in Cancer Prevention and Treatment," *Alternative Medicine Alert 8*, no. 10 (2005): 109–113.

Part III: Water

1. Columbia News Service, "The Pill vs. the Environment," *Chicago Tribune* (March 1, 2008), http://articles. chicagotribune.com/2008-03-01/news/0802290478_1_ birth-control-pills-hormones-synthetic-estrogen.

2. Robert D. Morris et al., "Chlorination, Chlorination By-products, and Cancer: A Meta-analysis," *American Journal of Public Health 82*, no. 7 (1992): 955–963.

3. K. P. Cantor et al., "Bladder Cancer, Drinking Water Source, and Tap Water Consumption: A Case-Control Study," *Journal of the National Cancer Institute 79*, no. 6 (1987): 1269–1279.

4. T. Ding et al., "The Relationship between Low Levels of Urine Fluoride on Children's Intelligence, Dental Fluorosis in Endemic Fluorosis Areas in Hulunbuir, Inner Mongolia, China," *Journal of Hazardous Materials 182*, nos. 2–3 (2011): 1942–1946.

5. Quanyong Xiang et al., "Analysis of Children's Serum Fluoride Levels in Relation to Intelligence Scores in a High and Low Fluoride Water Village in China," *Fluoride 44*, no. 4 (2011): 191–194.

6. Jennifer Anne Luke, "The Effect of Fluoride on the Pineal Gland," excerpts from a dissertation submitted to the School of Biological Sciences, University of Surrey, England (1997), www.infowars.com/the-effect-of-fluoride-on-the-pineal-gland.

7. S. C. Freni, "Exposure to High Fluoride Concentrations in Drinking Water Is Associated with Decreased Birth Rates," *Journal of Toxicology and Environmental Health 42*, no. 1 (1994): 109–121.

8. A. K. Susheela and P. Jethanandani, "Circulating Testosterone Levels in Skeletal Fluorosis Patients," *Journal of Toxicology and Clinical Toxicology 34*, no. 2 (1996): 183–189.

9. Committee on Fluoride in Drinking Water, National Research Council, *Fluoride in Drinking Water: A Scientific Review of EPA's Standards* (Atlanta, GA: The National Academies Press, 2006), 222.
10. Edward R. Schlesinger et al., "Newburgh-Kingston Caries-Fluorine Study. XIII. Pediatric Findings after Ten Years," *Journal of the American Dental Association 52*, no. 3 (1956): 296–306.
11. P. P. Bachinskii et al., "Action of the Body Fluorine of Healthy Persons and Thyroidopathy Patients on the Function of Hypophyseal-Thyroid System" (article in Russian), *Problemy Endokrinologii (Moskva) 31*, no. 6 (1985): 25–29.
12. George L. Waldbott, Albert W. Burgstahler, and H. Lewis Mckinney, *Fluoridation: The Great Dilemma* (Lawrence, KS: Coronado Press, 1979).
13. US Department of Health and Human Services, *Review of Fluoride: Benefits and Risks*, Report of the Ad Hoc Committee on Fluoride, Committee to Coordinate Environmental Health and Related Programs (1991).
14. Statistic Brain, "Bottled Water Statistics" (February 24, 2012), www.statisticbrain.com/bottled-water-statistics.
15. Arden Jobling-Hey, "Plastic Water Bottles and the Environment: How Bad Is Bad?" (August 21, 2012), www.bizenergy.ca/blog/plastic-water-bottles-and-the-environment-how-bad-is-bad.

Part IV: Earth's Gifts

1. Tina Bringslimark, Terry Hartig, and Grete Grindal Patil, "Psychological Benefits of Indoor Plants in Workplaces: Putting Experimental Results into Context," *Horticultural Science 42*, no. 3 (2007): 581–587.
2. Virginia I. Lohr and Caroline H. Pearson-Mims, "Interior Plants May Improve Worker Productivity and Reduce Stress

in a Windowless Environment," *Journal of Environmental Horticulture 14*, no. 2 (1996): 97–100.

3. S. Park and R. H. Mattson, "Ornamental Indoor Plants in Hospital Rooms Enhanced Health Outcomes of Patients Recovering from Surgery," *Journal of Alternative and Complementary Medicine 15*, no. 9 (2009): 975–980.

4. S. Park and R. H. Mattson, "Therapeutic Influences of Plants in Hospital Rooms on Surgical Recovery," *Horticultural Science 44*, no. 1 (2009): 102–105.

5. Joan L. Bardach, *Some Principles of Horticultural Therapy with the Physically Disabled* (Mt. Vernon, VA: National Council for Therapy and Rehabilitation through Horticulture, 1975).

6. Cammie K. Coleman and Richard H. Mattson, "Influences of Foliage Plants on Human Stress during Thermal Biofeedback Training," *Horticultural Technology 5*, no. 2 (1995): 137–140.

Suggested Reading

Part I: Air

Aragon, Daleen, Carla Farris, and Jacqueline F. Byers. "The Effects of Harp Music in Vascular and Thoracic Surgical Patients." *AlternativeTherapies in Health and Medicine 8*, no. 5 (2002): 52–54, 56–60.

Benson, Jonathan. "California Desert City Proposes Broadcasting Fake Bird Sounds to Cheer Up Residents." *Natural News* (February 5, 2011), www.naturalnews.com/031219_bird_sounds_California.html.

Block, Steven M., Debby Jennings, and Lisa R. David. "Live Harp Music Decreases Salivary Cortisol Levels in Convalescent Premature Infants." *Pediatric Research 53*, no. 4 (2003): 469A–470A.

Briggs, Tami. "Live Harp Music Reduces Anxiety of Patients Hospitalized with Cancer." *The Harp Therapy Journal 8*, no. 4 (Winter 2003–2004).

Byers, Jacqueline F, and K. A. Smyth. "Effect of Music Intervention on Noise Annoyance, Heart Rate, and Blood Pressure in Cardiac Surgery Patients." *American Journal of Critical Care 6*, no. 3 (1997): 183–191.

Caine, Janel. "The Effects of Music on the Selected Stress Behaviors, Weight, Caloric and Formula Intake, and Length of Hospital Stay of Premature and Low Birth Weight Neonates in a Newborn Intensive Care Unit." *Journal of Music Therapy 28*, no. 4 (1991): 180–192.

Cameron, Layne. *"New Scientific Field to Study Ecological Importance of Sounds." Michigan State University News* (March 8, 2011), http://news.msu.edu/story/9029.

Eisele, Melodye. "Bringing Heart to the Hospital: Special Neonatal Care Unit Benefits from Harp Music." *The Harp Therapy Journal* (Summer 1999): 10–11.

Elliott, Dave. "The Effects of Music and Muscle Relaxation on Patient Anxiety in a Coronary Care Unit." *Heart and Lung 23*, no. 1 (1994): 27–35.

Good, Marion et al. "Use of Relaxation and Music to Reduce Post-Surgical Pain." *Journal of Advanced Nursing 33*, no. 2 (2001): 208–215.

Gutierrez, D. "Bird Songs Lend Therapeutic Powers to Hospital Patients." *Natural News* (December 4, 2010), www.naturalnews.com/030612_bird_songs_therapy.html.

Kuo, Frances E. and Andrea F. Taylor. "A Potential Natural Treatment for Attention-Deficit/Hyperactivity Disorder: Evidence from a National Study." *American Journal of Public Health 94*, no. 9 (2004): 1580–1586.

National Park Services, U.S. Department of the Interior. "Natural Sounds and the Night Skies." (Last updated, April 16, 2012), www.nature.nps.gov/sound_night.

National Park Service, U.S. Department of the Interior. "Sounds We Protect." (Last updated, April 3, 2012, www.nature.nps.gov/naturalsounds/natural/index.cfm#nullOutside/Be-Out-There/Why-Be-Out-There/Benefits.aspx.

Williams, Sarajane. "Patients with Parkinson's Disease Find Relief with Harp Music." *The Harp Therapy Journal* (Spring 2001): 6–7.

Wolverton, Bill C. *How to Grow Fresh Air: 50 Household Plants That Purify Your Home and Office.* New York: Penguin Books, 1997.

Part II: Light

Hobday, Richard. *The Healing Sun: Sunlight and Health in the 21st Century.* Forres, Scotland: Findhorn Press, 1999.

Lieberman, Jacob. *Light: Medicine of the Future: How We Can Use It to Heal Ourselves Now.* Santa Fe, NM: Bear & Company, 1991.

Kime, Zane R. *Sunlight.* Pengrove, CA: World Health Publications, 1980.

Part III: Water

Kellar, Casey. *Natural Cleaning for Your Home: 95 Pure and Simple Recipes*. Asheville, NC: Lark Books, 1998.

Part IV: Earth's Gifts

Lembo, Margaret Ann. *Essential Guide to Crystals, Minerals and Stones*. Woodbury, MN: Llewellyn Publications, 2013.

Hall, Judy. *The Crystal Bible 2*. London, UK: Octopus Publishing Group, 2009.

Pierce, Linda Breen. *Choosing Simplicity: Real People Finding Peace and Fulfillment in a Complex World*. Seattle, WA: Gallagher Press, 2000.

Prattis, Ian, ed. *Earth My Body, Water My Blood*. Ottawa, ON: Baico Publishing, 2011.

Jefferies, Tash. *The Little Book of Green Minutes: Simple Steps to Health, Happiness, and Empowered Living*. Toronto, ON: Expert Author Publishing, 2012.

Resources

Part I: Air

Air Diffusers
- Abundant Health
  www.abundanthealth4u.com
- Diffuser World
  www.diffuserworld.com
- Young Living
  www.youngliving.com

Essential Oils
- Raise Your Vibration
  http://raiseyourvibration.ca/healthy-home-healthy-you

Harp Music
- Harp.com
  www.harp.com
- The Harp Connection
  www.harpconnection.com

Nature Sounds
- Best Nature Sounds
  www.bestsoundsofnature.com

- Listening Earth
  www.listeningearth.com
- Nature Sounds.ca
  http://nature-cd-store.naturesounds.ca

Sound and Brainwaves
- EOC Institute
  www.eocinstitute.org
- Hemi Sync
  www.hemi-sync.com

Wind Chimes
- Amish Handcrafted Wind Chimes
  www.amishhandcraftedmetalwindchimes.com

Part II: Light

Cappings Beeswax Candles
- The Original Pheylonian Beeswax Candles (contains a directory of where to purchase these candles in Canada, the Unites States
  and Internationally
  http://naturalbeeswaxcandles.com

Citizens for Safe Technology
United States and International
- www.citizensforsafetechnology.com

Canada
- http://citizensforsafetechnology.org

Part III: Water

Pure Water Solutions
- MineralPRO (sells remineralized reverse osmosis drinking water systems)
  www.mineralpro.com

Green Cleaning Products
- Raise Your Vibration
  http://raiseyourvibration.ca/healthy-home-healthy-you

United States
- http://greencleaningproductsllc.com/commercial-green-clean
- www.terafore.net

Canada
- http://effeclean.com/about-us
- www.naturalproductsna.com/commercial.html

Part IV: Earth's Gifts

Organic Gardening
- Garden Organic
  www.gardenorganic.org.uk
- Forum for the Future: Action for a Sustainable World
  www.forumforthefuture.org.uk

Community Gardens
- American Community Gardening Association (Lists community gardens in Canada and the United States)
  www.communitygarden.org

Heirloom and Organic Seeds

- Heirloom Seeds (Info about heirloom produce and gardening; ship only in the United States)
www.heirloomseeds.com
- Hawthorn Farm Organic Seeds (Organic vegetable, flower, herb an grass seeds; ship only in Canada)
www.hawthornfarm.ca
- Terra Edibles (Organically grown heirloom seeds and other products)
www.terraedibles.ca

Eco-Friendly Living

- Ethical Ocean (online retailer of ethical goods from around the globe)
www.ethicalocean.com
- The Laundry Tarts (Maker of natural laundry detergents, available in Canada, the United States and Europe)
www.thelaundrytarts.com

About the Author

*D*awn James is the founder of raiseyourvibration.ca, an organization dedicated to providing education, inspiration and support related to understanding and enhancing vibrational frequency for overall health and well-being (personally and globally). Dawn became a sound healer and writer following a series of spiritual events in 2003 that opened her eyes and heart to the world of spirituality, higher consciousness and vibrational frequency. In that moment, she realized her soul purpose to usher in a new world, where peace and harmony would be the norm. She accepted her new role as "teacher."

Today she shares her knowledge and gift of healing through sacred sound circles, workshops, home-study programs, radio, TV and retreats. She is an international speaker, green-living advocate, musician and author of several books on vibrational frequency.

Learn more at www.raiseyourvibration.ca.

Also by Dawn James

Raise Your Vibration, Transform Your Life: A Practical Guide for Attaining Health, Vitality and Inner Peace (Paperback and eBook)

Eleva Tu Vibración, Transforma Tu Vida: Una Guía Práctica para Obtener Mejor Salud, Vitalidad y Paz Interior (Spanish eBook)

I hope you have enjoyed *How to Raise the Vibration around You: Volume I: Working with the 4 Elements to Create Healthy and Harmonious Living Spaces*. This is the second book of a trilogy on raising the vibrational frequencies within and around you. Autographed copies are available at www.raiseyourvibration.ca. The first book, *Raise Your Vibration, Transform Your Life: A Practical Guide for Attaining Better Health, Vitality and Inner Peace*, is also available and sold globally online by various book retailers.

Let's Stay Connected

Join me on the journey of becoming a harmonic being, and join our online community at **www.raiseyourvibration.ca**, a site dedicated to providing education, inspiration and support related to understanding and enhancing vibrational frequency for your overall well-being.